GW00384520

Mr C. F. Gardiner
40a Polsloe Road
Exeter EX1 2DN.
Springer Pri. 26/6/98.

Universitext

Editorial Board
(North America)

S. Axler
F.W. Gehring
K.A. Ribet

Springer

New York
Berlin
Heidelberg
Barcelona
Budapest
Hong Kong
London
Milan
Paris
Santa Clara
Singapore
Tokyo

Universitext

Editors (North America): S. Axler, F.W. Gehring, and K.A. Ribet

Aksoy/Khamsi: Nonstandard Methods in Fixed Point Theory
Andersson: Topics in Complex Analysis
Aupetit: A Primer on Spectral Theory
Booss/Bleecker: Topology and Analysis
Borkar: Probability Theory: An Advanced Course
Carleson/Gamelin: Complex Dynamics
Cecil: Lie Sphere Geometry: With Applications to Submanifolds
Chae: Lebesgue Integration (2nd ed.)
Charlap: Bieberbach Groups and Flat Manifolds
Chern: Complex Manifolds Without Potential Theory
Cohn: A Classical Invitation to Algebraic Numbers and Class Fields
Curtis: Abstract Linear Algebra
Curtis: Matrix Groups
DiBenedetto: Degenerate Parabolic Equations
Dimca: Singularities and Topology of Hypersurfaces
Edwards: A Formal Background to Mathematics I a/b
Edwards: A Formal Background to Mathematics II a/b
Foulds: Graph Theory Applications
Friedman: Algebraic Surfaces and Holomorphic Vector Bundles
Fuhrmann: A Polynomial Approach to Linear Algebra
Gardiner: A First Course in Group Theory
Gårding/Tambour: Algebra for Computer Science
Goldblatt: Orthogonality and Spacetime Geometry
Gustafson/Rao: Numerical Range: The Field of Values of Linear Operators and Matrices
Hahn: Quadratic Algebras, Clifford Algebras, and Arithmetic Witt Groups
Holmgren: A First Course in Discrete Dynamical Systems
Howe/Tan: Non-Abelian Harmonic Analysis: Applications of $SL(2, R)$
Howes: Modern Analysis and Topology
Humi/Miller: Second Course in Ordinary Differential Equations
Hurwitz/Kritikos: Lectures on Number Theory
Jennings: Modern Geometry with Applications
Jones/Morris/Pearson: Abstract Algebra and Famous Impossibilities
Kannan/Krueger: Advanced Analysis
Kelly/Matthews: The Non-Euclidean Hyperbolic Plane
Kostrikin: Introduction to Algebra
Luecking/Rubel: Complex Analysis: A Functional Analysis Approach
MacLane/Moerdijk: Sheaves in Geometry and Logic
Marcus: Number Fields
McCarthy: Introduction to Arithmetical Functions
Meyer: Essential Mathematics for Applied Fields
Mines/Richman/Ruitenburg: A Course in Constructive Algebra
Moise: Introductory Problems Course in Analysis and Topology
Morris: Introduction to Game Theory
Polster: A Geometrical Picture Book
Porter/Woods: Extensions and Absolutes of Hausdorff Spaces
Ramsay/Richtmyer: Introduction to Hyperbolic Geometry
Reisel: Elementary Theory of Metric Spaces
Rickart: Natural Function Algebras

(continued after index)

Burkard Polster

A Geometrical Picture Book

With 405 Illustrations

 Springer

Burkard Polster
Department of Pure Mathematics
The University of Adelaide
Adelaide, SA 5005
Australia

Editorial Board (*North America*):

S. Axler
Mathematics Department
San Francisco State
  University
San Francisco, CA 94132
USA

F.W. Gehring
Mathematics Department
East Hall
University of Michigan
Ann Arbor, MI 48109
USA

K.A. Ribet
Department of Mathematics
University of California
  at Berkeley
Berkeley, CA 94720-3840
USA

Mathematics Subject Classification (1991): 05Bxx, 51-01, 51-02

Library of Congress Cataloging-in-Publication Data
Polster, Burkard.
    A geometrical picture book / Burkard Polster.
        p.   cm. — (Universitext)
    Includes bibliographical references and index.
    ISBN 0-387-98437-2 (hardcover : alk. paper)
    1. Geometry—Pictorial works.   I. Title.
QA447.P64   1998
516′.12′0222—dc21   97-48854

Printed on acid-free paper.

© 1998 Springer-Verlag New York, Inc.
All rights reserved. This work may not be translated or copied in whole or in part without the written permission of the publisher (Springer-Verlag New York, Inc., 175 Fifth Avenue, New York, NY 10010, USA), except for brief excerpts in connection with reviews or scholarly analysis. Use in connection with any form of information storage and retrieval, electronic adaptation, computer software, or by similar or dissimilar methodology now known or hereafter developed is forbidden.
The use of general descriptive names, trade names, trademarks, etc., in this publication, even if the former are not especially identified, is not to be taken as a sign that such names, as understood by the Trade Marks and Merchandise Marks Act, may accordingly be used freely by anyone.

Production managed by Karina Mikhli; manufacturing supervised by Jeffrey Taub.
Camera-ready copy prepared from the author's LaTeX files.
Printed and bound by R.R. Donnelley and Sons, Harrisonburg, VA.
Printed in the United States of America.

9 8 7 6 5 4 3 2 1

ISBN 0-387-98437-2 Springer-Verlag New York Berlin Heidelberg   SPIN 10659738

To my parents

Introduction

What This Book Is All About

Like most people, you probably glanced through the book before you started reading this introduction. If this is so, you have already discovered what this book is all about, namely pictures, lots and lots of pictures of different kinds of geometrical structures.

Our main goal in writing this book was to provide a source for aesthetically pleasing two- and three-dimensional models of some of the most fundamental incidence geometries students, teachers, and researchers come across in the theories of codes, designs, graphs, groups, statistical designs, and, of course, incidence geometry itself. Examples of such geometries are affine and projective planes and spaces, linear spaces, (symmetric) designs, circle planes, and generalized polygons, to name just a few. Such a source book will prove especially useful in teaching undergraduate and beginning graduate courses dealing with these geometries and in conveying some of the beauty of the objects geometers are obsessed with to the general mathematical and nonmathematical public.

The book is divided into two main parts. The first one presents pictures of finite geometries with small numbers of points, the second part pictures of topological geometries, mainly those that live on surfaces.

There are a number of excellent, highly accessible introductions to the various kinds of geometries we are dealing with. This is especially true for finite geometries. Although we have tried to make this book as self-contained as possible, our aim is to complement rather than better these introductions. Accordingly, we keep theory to a minimum and refer to these introductions for details, proofs, etc., of the facts used in this book. We recommend the following textbooks and survey articles: [30], [31], [56], [90], and [104] for affine and projective planes; [7], [24], [45], and [57] for design theory; [49] for configurations; [21], [50], [51], and [52] for everything else in the known finite universe; [42], [90], and [103] for geometries on surfaces.

In the first part of the book we very much concentrate on drawing well what can be drawn rather than trying to present pictures that illuminate every single aspect of finite geometry. There are some natural limitations to which pictures can and should be drawn. Some of the most important factors here are the relative importance of the geometry that is to be portrayed, the number of points and lines involved, and the presence or absence of symmetries that can be used to produce an appealing picture of a geometry. Also, there are many wonderful things that happen in geometries of small orders that can be turned into attractive pictures and that do not have counterparts for higher orders. On the other hand, for small orders all the geometries tend to be classical. For example, the smallest nonclassical affine plane has 81 points and is just a little bit too large for a picture of it to be included in this book. These factors do not necessarily have anything to do with each other and produce a somewhat distorted picture of what finite geometry is all about. To get a more balanced and well-rounded picture of finite geometry, you are better off using this book side by side with some of the textbooks and articles mentioned above.

For a number of fields that are part of finite geometry, textbooks are available that make very good use of pictures. For matroids there is the excellent book by Oxley [72], and for a complete list of the linear spaces with up to 9 points see [2, Appendix]. The interested reader is referred to these books for pictures of matroids and linear spaces.

In the second part of the book we are dealing with topological geometries and in particular with geometries living on surfaces. Just as in the finite case, these geometries can be grouped around certain classical examples, which are in this case associated with the real numbers. The major players here are the flat affine and projective planes grouped around the real projective plane, and flat circle planes grouped around the classical Möbius, Laguerre, and Minkowski planes over the reals. All these geometries have infinite point and line sets, and pictures of these geometries will usually include only the surface they are living on, a couple of sample lines, and a rule, usually the action of some group, that tells you how to generate all the other lines from the given ones. Unlike in the finite case, we will arrive at a pretty well-rounded overall picture of what topological geometries on surfaces are all about, in that we can draw meaningful pictures of virtually all important examples of such geometries. In particular, it is possible to present some beautiful constructions of nonclassical examples of such geometries.

Finally, we want to point out that a number of topics that we do not know much about were omitted from this book. On the other hand, we have included some topics that might not be regarded as belonging to incidence geometry by some colleagues. All this should not be regarded as judgment on our part as to which topics are "good" and "bad" geometry. Anyway, this is a very personal selection of topics. We hope that you enjoy this book as much as we enjoyed writing and drawing it.

Rough Description of Contents

Here is a rough description of the contents of this book. To the expert a quick look at the table of contents should suffice to get a clear picture of what this book is all about.

In the "finite" part of the book we first give a general introduction to geometries by having a close look at the Fano plane in Chapter 1. The Fano plane is the smallest projective plane and one of the major players in this part of the book. We present pictures of this plane that illustrate some important principles for constructing the pictures in this book. We draw attention to the different kinds of objects and substructures that can be represented by diagrams.

The introductory section is followed by a section on designs that pop up in various places in the book. It serves as both a rough introduction to designs and an index to the locations of symmetric and other kinds of designs in the book. In this section we also present the one-point extension of the Fano plane and say something about the generalization of this construction to Hadamard 2-designs. Finally, we present pictures of the smallest Steiner triple systems and a pictorial solution to Kirkman's schoolgirl problem.

In Chapter 3 we collect some facts about plane configurations and their abstract counterparts. We pay special attention to configurations with three points on every line such as the Fano, Desargues, and Pappus configurations. We conclude this chapter by having a quick look at tree-planting puzzles, which are closely related to plane configurations and whose solutions often provide us with good raw material that can be turned into nice pictures of affine planes.

In Chapter 4 we give an introduction to generalized quadrangles. We first concentrate on the smallest (nontrivial) generalized quadrangle. This geometry proves to be of equal importance to us as the Fano plane, as it can be embedded, just like the Fano plane, in a host of geometries with small parameters. We present a two-dimensional and two three-dimensional models of this geometry and show how such important concepts as spreads, ovoids, and polarities show up in these models. We also extend models of this generalized quadrangle to models of the generalized quadrangles of orders $(2, 4)$ and $(4, 2)$. Some of the most important construction principles employed to construct the pictures in this book pop up for the first time in this chapter. Also contained in this chapter is a detailed introduction to the different techniques and tricks for viewing the stereographic images in this book.

Chapter 5 deals with three nice representations of the smallest projective space, which again extend our favourite representations of the smallest nontrivial generalized quadrangle. Among other things, we present a complete list of all spreads of this space and an example of a packing of spreads that corresponds to a solution of Kirkman's schoolgirl problem.

In Chapter 6 we describe a number of plane and spatial models of the affine and projective planes of order 3. We also give a "semipictorial" description of the classical $5 - (12, 6, 1)$ design as an extension of the affine plane of order 3.

In Chapter 7 we concentrate on the projective plane of order 4, which is one of the most important geometries in finite geometry. We again use the fact that the smallest nontrivial generalized quadrangle is embedded in this plane to rebuild it around our favourite representations of this generalized quadrangle. We also rebuild it around a unital and present pictures of partitions of this plane into three Fano subplanes.

In Chapter 8 we present a pictorial version of Beutelspacher's construction of the projective plane of order 5 built around an oval, and a spatial model of this plane on the dodecahedron. We also have another look at the Desargues configuration as the geometry of interior points and exterior lines of an oval in the projective plane of order 5.

In Chapter 9 we present a series of plane models, called star diagrams, of the classical affine planes up to order 8. These pictures were introduced by Breach in [15]. We also exhibit a nonclassical oval and a Fano configuration right in the middle of the star diagram of the affine plane of order 8.

In Chapters 10 and 11 we have a closer look at a straightforward generalization of projective planes, the so-called biplanes and semibiplanes. We present pictures of the smallest of these planes and their associated Hussain graphs.

In Chapter 12 we present a unifying definition of Benz (circle) planes together with some models of the Möbius, Laguerre, and Minkowski planes of orders 2 and 3. Probably the nicest model in this section is a representation of the Möbius plane of order 3 on the tetrahedron such that all symmetries of the tetrahedron are automorphisms of the Möbius plane.

Chapter 13 deals with generalized polygons, in particular generalized hexagons, octagons, and 12-gons. The incidence graphs of projective planes and generalized quadrangles provide us with some nice examples of such geometries. Also included in this section are diagrams by A. Schroth that depict the two classical generalized hexagons of order $(2, 2)$ and a list of the smallest cages, some of the most important of which are generalized n-gons.

In the following chapter we describe how to build models of the three-dimensional representations of geometries in this book from pipe cleaners.

In the final chapter of this part of the book we present a number of games and puzzles such as the popular card game "Set" that are based on important geometries discussed in this book.

The second part of the book deals with geometries whose point sets are surfaces and whose lines or circles are "nice" curves drawn on these surfaces.

In Chapter 16 we give an introduction to geometries on surfaces and how such geometries can be represented in pictures by looking at examples of flat affine planes, that is, close relatives of the Euclidean plane.

Chapter 17 gives an overview of the most important kinds of geometries on surfaces that have topological circles as lines.

Chapter 18 deals with flat projective planes, that is, close relatives of the projective plane over the real numbers. The flat projective planes together with their affine counterparts are those among the geometries on surfaces that are best understood.

Chapters 19, 20, and 21 are structured in a similar way. They deal with the different kinds of flat circle planes of rank 3. We construct some of the most important examples and representations of such circle planes, describe characterizations of the classical planes in terms of the Miquel, bundle, and rectangle configurations, and describe various subgeometries and Lie geometries associated with such circle planes.

Appendix A contains an index of the various models of geometries on regular solids described in this book.

There are about thirty stereographic images contained in the main part of the book that can be viewed with both the parallel and the cross-eyed techniques. Appendix B contains a complete list of all the stereograms that have been modified such that they can be viewed with the help of a pocket mirror.

Pictures

Most of the pictures in this book were created with the program xfig. The stereograms were produced with the programs MetaPost and Mathematica and the colour inserts with the programs Adobe Illustrator and Adobe Photoshop.

When flipping through many textbooks with pictures, one often gets the impression that all of them look the same. Although the single diagrams may well look very professional and well balanced, the uniformity in the way they are drawn makes them rather boring to look at.

We have tried to keep our pictures as nonuniform and "nonboring" as possible by varying features of the diagrams such as overall size, thickness of lines, and arrangements of blocks of pictures.

Acknowledgments

Apart from always having been obsessed with the need to draw pictures, the main incentives for writing this book were Derrick Breach's presenting me with a reprint of his article [15] on his beautiful star diagrams of affine planes, Alan Offer's showing me his wonderful pipe-cleaner model of the smallest three-dimensional projective space, and Andreas Schroth's amazing diagrams of the two generalized hexagons of order $(2, 2)$.

The pictures of the generalized hexagons in Section 13.3 and the stereograms of $PG(3, 2)$ and the generalized quadrangle in Sections 4.1.2 and 5.2.1 were created by Andreas Schroth using the program MetaPost. Special thanks to him for letting me use his pictures in this book.

Many other colleagues and friends have pointed out beautiful constructions that can be turned into pictures, have read or looked at parts of the book, and offered helpful advice. In particular, thanks are due to Susan Barwick, Albrecht Beutelspacher, Dieter Betten, Aart Blokhuis, David Cook, Alex Cowie, Rainer Löwen, Paul McCann, Christine O'Keefe, Tim Penttila, Gordon Royle, Tze-lan Sang, Günter Steinke, and Chris Tuffley.

Furthermore, I wish to thank the Australian Research Council for its assistance.

Finally, and most importantly, I would like to thank Karl Strambach for his unrivalled support over the years.

Burkard Polster
Adelaide, November 1997

Contents

Part I

Finite Geometries

1
Introduction via the Fano Plane

1.1 Geometries—Basic Facts and Conventions

In this book a *geometry* will usually consist of a nonempty *point set* and a nonempty *line set*. Here a *line* is a subset of the point set containing at least two points and a point is contained in at least two lines. For example, the Euclidean plane is a geometry whose point set is the xy-plane and whose lines are the straight lines. Sometimes, for historical reasons, lines will also be called *blocks* or *circles*. For example, in the geometry of circles on a sphere the lines are called circles. Usually our geometries also satisfy a number of *axioms*. For example, both the Euclidean plane and the geometry of circles satisfy an "axiom of joining"; that is, two points in the Euclidean plane are always contained in a unique line, and three points on a sphere are always contained in exactly one circle. In fact, various versions of the axiom of joining are the most important kind of axioms that we will come across, and virtually all the geometries considered in this book satisfy an axiom of joining.

A geometry is called *finite* or *infinite* depending on whether its point set contains a finite or infinite number of points.

A *line pencil* associated with a set of one or more points of a geometry is the set of all lines containing all the points in the set. The points in the set are called the *carriers* of the pencil. Two or more points are *collinear* if they are all contained in some line.

Two lines of a geometry are called *parallel* if they coincide or if they do not have any points in common. Two points of a geometry are called *parallel* if they coincide or if they are not connected by a line. If being parallel defines an equivalence relation on the point set or on the line set, then the equivalence classes of this equivalence relation are called *parallel classes*. If being parallel on the point and the line sets both define equivalence relations, then the geometry is called *divisible*.

Given a point p in a geometry, the *derived geometry at the point* consists of all the points different from p. Associated with every line of the original geometry that contains p there is a line of the derived geometry that contains all the points of the line except p.

An *isomorphism* between two geometries is a one-to-one map from the point set of the first geometry to the point set of the second geometry, that induces a one-to-one map between the line sets. An *automorphism* of a geometry is an isomorphism from the geometry to itself. The set of automorphisms of a geometry forms a group, and the theory of groups and, in particular, permutation groups provides a powerful tool for investigating geometries.

Sometimes geometries will be given in a more abstract form, where the point and line sets are disjoint sets, together with an *incidence relation*, that is, a rule that prescribes when a point is contained in (= incident with) an (abstract) line. A geometry given in this abstract form can always be turned into the more elementary form by replacing an abstract line with the set of points incident with it. Consider a geometry given in the elementary form. Call the points Lines and the lines Points and let a Point be incident with a Line if the Line (an old point) is contained in the Point (an old line). This geometry is called the *dual* of the original geometry.

Example: Let us start with the Euclidean plane from which the vertical lines have been removed. This geometry is *self-dual*. That is, after turning the dual of this geometry back into its elementary form, we end up with the geometry we started out with.

Graphs

A *graph* is a geometry in which every line contains exactly two points. Points of such a geometry are usually called *vertices* and lines *edges*. Note that, contrary to the common usage of the word graph, our graphs never contain "loops" or "multiple edges." A *path* of length i is a sequence of $i+1$ vertices such that consecutive vertices are connected by an edge. A path is *closed* if its first and last vertex coincide. A path is *simply closed* if it is closed and if among its vertices only the first and the last coincide. A path *connects* its first with its last vertex. A graph is *connected* if any two of its vertices can be connected by a path. The distance between two vertices is the length of the shortest path connecting them. The *diameter* of a graph is the maximum *distance* between two of its vertices. The *girth* of a graph is the length of its shortest simply closed path. A graph is called *regular* with *valency k* if every vertex is contained in exactly *k* edges. A graph is called *complete* if any two of its vertices are connected by an edge. A graph is called *bipartite* if its vertices can be coloured with two colours such that the two vertices contained in every edge have different colours.

Example: The dodecahedron can be viewed as a graph with 20 vertices and 30 edges. It is a connected, regular graph of valency 3. Its diameter and its girth are both 5. It is not bipartite.

The complete information about any geometry is contained in its *incidence graph*. This graph is constructed as follows: The vertices of the graph are the points and lines of the geometry, and two vertices of the graph are connected by an edge if and only if they correspond to a point and a line of the geometry such that the point is contained in the line. A geometry is *connected* if its incidence graph is connected.

1.2   Projective Planes

In this section we consider the smallest projective plane and, using it as a guiding example, introduce many concepts that play an important role in the study of geometries in general. A good reference for most of the facts mentioned in this section is [56].

A projective plane is a geometry that satisfies the following three axioms:

Axioms for projective planes

(P1) Two distinct points are contained in a unique line.

(P2) Two distinct lines intersect in a unique point.

(P3) There exist four points of which no three are incident with the same line.

Here is a picture of such a projective plane, the so-called *Fano plane*.

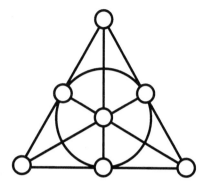

Anybody who claims to know anything about geometry is familiar with this picture. In fact, given the information that a book on geometry contains any pictures at all, it is a safe bet that this picture of the Fano plane is one of them.

Note that there are 7 points and 7 lines (the circle counts as a line!). Note also that in this geometry every line contains 3 points and every point is contained in 3 lines. It is an easy exercise to check that the geometry really satisfies the three axioms above. To check Axiom P1, have a quick look at the following three essentially different line pencils through points of the plane.

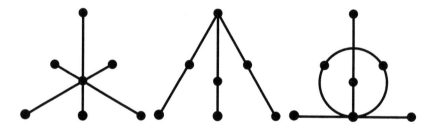

Every such line pencil through a point partitions the other points of the plane. This implies that the first axiom is satisfied.

The Fano plane, like most geometries depicted in this book, is a *homogeneous geometry*. This means that two different points of the geometry are indistinguishable, or equivalently, that there is an automorphism of the geometry that carries one point to the other. This also implies that two line pencils are indistinguishable and only appear to be different in pictures.

We recall some well-known consequences of the above axioms for *finite projective planes*. Let $\mathcal{P}$ be such a plane.

Then there is an integer $n > 1$, called the *order* of the projective plane $\mathcal{P}$, such that every line contains $n + 1$ points, every point is contained in $n + 1$

lines, and both its point set and its line set contain $n^2 + n + 1$ elements. This means that the Fano plane is a projective plane of order 2.

The *classical examples* of projective planes are the projective planes over fields. In a well-known construction of a model of such a classical example over a field K, the point set is the set of all one-dimensional and the line set the set of all two-dimensional subspaces of the three-dimensional vector space over K. In this model the three axioms are easily verified. For example, Axiom P1 translates into the well-known fact that two one-dimensional subspaces of a three-dimensional vector space are contained in a uniquely determined two-dimensional subspace.

Given any prime power n, there is a unique field containing exactly n elements, and the associated finite projective plane has order n. This projective plane is usually denoted by $PG(2, n)$ (PG stands for "projective geometry," the 2 stands for "two-dimensional"). The Fano plane is the projective plane associated with the field of order 2. In fact, it is unique up to isomorphism and is therefore also referred to as the projective plane of order 2. The same is true for the projective planes of orders 3, 4, 5, 7, and 8. There are examples of nonclassical projective planes of order 9 and many of higher order. Many geometers believe that all finite projective planes have prime-power orders just like the classical examples. It is known, for example, that there are no finite projective planes of orders 6 and 10. Infinitely many further orders are excluded by the famous *Bruck–Ryser theorem* (see [56]).

Projective planes are geometries that are very rich in structure, and especially the classical examples are highly symmetrical geometries. We remark that Axiom P3 excludes some geometries that are "poor" in structure but satisfy the first two axioms. Here are three pictures of geometries that are excluded by Axiom P3.

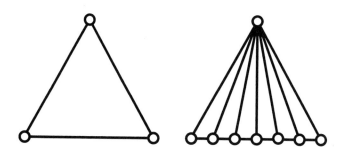

1.3 Affine Planes

By removing a line and all the points contained in it from a projective plane, we arrive at a geometry that satisfies the following axioms:

Axioms for affine planes

(A1) Two distinct points are contained in a unique line.

(A2) Given a line and a point, there is a unique line through the point that is parallel to the given line.

(A3) There exist three points that are not all contained in a line.

Every geometry that satisfies these axioms is called an *affine plane*.

The affine plane we arrive at by removing a line from the *real projective plane*, that is, the classical projective plane associated with the real numbers, is the Euclidean plane. All affine planes obtained from a classical projective plane by removing a line are isomorphic, and every affine plane has a unique *projective extension* to a projective plane. By removing different lines in the traditional picture of the Fano plane, we arrive at the following two representations of the affine plane associated with it.

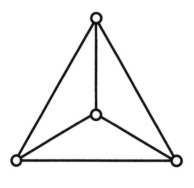
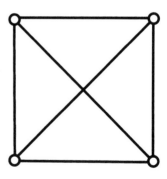

We arrive at the first picture by removing the "circle" and at the second picture by removing one of the sides of the outer triangle and then straightening out the resulting picture.

Note that both pictures are projections of the tetrahedron onto the plane. This means that the geometry whose points and lines are the vertices and edges of the tetrahedron, respectively, is another model of this affine plane.

An affine plane is said to be of order n if it is derived from a projective plane of order n. As in the Euclidean plane, the line set of every affine plane is partitioned into *parallel classes*, that is, sets of parallel lines. By Axiom A2, given a parallel class and a point, there is exactly one line in the parallel class that contains the point. In the projective extension of an affine plane the lines in a parallel class are extended by a common point. In an affine plane of order n every line contains n points; every point is contained in $n+1$ lines; and there are n^2 points, n^2+n lines, $n+1$ parallel classes, and n lines in every parallel class. Here are the essentially different parallel classes of lines in our diagrams.

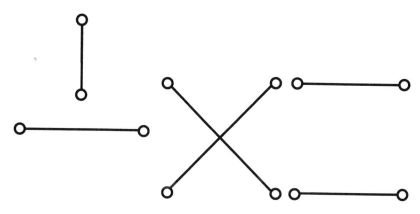

1.4   Automorphisms

In the pictures presented in this book we try to exhibit as many of the symmetries and automorphisms of geometries as possible. In the traditional picture of the Fano plane we see that a rotation of 120 degrees around the middle point of the diagram corresponds to an automorphism of the Fano plane of order 3. Also, reflections in the three symmetry axes of the diagram correspond to involutory automorphisms. This means that all the symmetries of the regular triangle that underlie the diagram correspond to automorphisms of our geometry.

Similarly, given a geometry that admits an automorphism of order m, we will often try to draw a picture of it on a regular m-gon such that all or most of the symmetries of the m-gon translate into automorphisms of the geometry. Of course, it is important to choose the right kind of automorphism if one has something like this in mind. As a general rule, good automorphisms are those that have few fixed points (preferably 0 or 1) and most of whose orbits have equal length m. Furthermore, the bigger m is, the better. In pictures that were generated in this way, it is also no longer necessary to draw in all lines, but only some *generators*. The remaining lines are then simply the images of these generators under suitable rotations. The main advantage of this approach is that the generator-only pictures appear much less crowded than those that include all possible lines. Here is a *generator-only* version of the traditional picture of the Fano plane that also happens to be one of the line pencils that we looked at before.

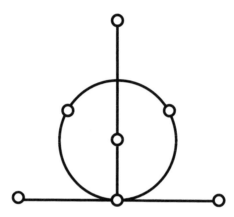

Here is another picture of the Fano plane that is based on an automorphism of order 3, with the generator-only picture on the left and the full picture on the right.

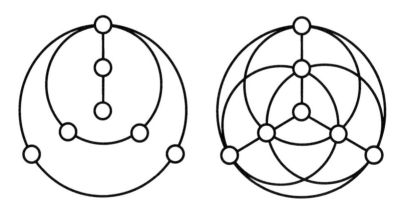

Later on we model the Fano plane on the tetrahedron such that all symmetries of the tetrahedron correspond to automorphisms of the Fano plane. We also model geometries on other highly symmetric graphs and objects such as the other four Platonic solids, that is, the cube, the octahedron, the icosahedron, and the dodecahedron.

1.5 Polarities

A *polarity* of a geometry is a function that maps the points to the lines and the lines to the points of the geometry, such that applying the function twice yields the identity and such that incident point/line pairs are mapped onto incident line/point pairs.

Here is a polarity of the Fano plane given in three generator-only diagrams. In all three cases the solid point is exchanged with the line.

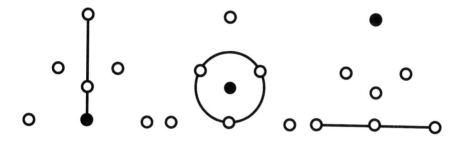

An *absolute* point of a polarity is a point that is contained in its image under the polarity. Similarly, an absolute line of a polarity is a line that contains its image under the polarity. The absolute points and lines of the above polarity are the centers of the edges of the outer triangle and the lines through the middle of the diagram, respectively. The sets of absolute points and lines of polarities of projective planes usually form very interesting sets. Consider a classical projective plane of order n. Then three different sets of absolute points of polarities are possible:

- The order of the plane is even, and the set contains $n + 1$ elements. Then the set is a nonabsolute line. This is what happens in the above example.

- The order is odd, and the set contains $n + 1$ elements. Then the set is a nondegenerate conic section, that is, one of the classical ovals in the plane. See the next section for more information about ovals.

- The order of the plane is a square, and the set contains $n^{3/2} + 1$ points. Then this set together with the set of nonabsolute lines forms a unital. We will encounter unitals in the chapter on the projective plane of order 4.

1.6   Ovals and Hyperovals

An oval in a projective plane is a nonempty set that satisfies the following two axioms:

Axioms for ovals in projective planes

(O1) No three points of the set are collinear.

(O2) Every point of the set is contained in a unique *tangent line*, that is, a line that intersects the set in only one point.

An oval in a projective plane of order n contains $n+1$ points. All classical projective planes contain ovals. The classical ovals are the so-called *non-degenerate conic sections* (see [56]). If n is even, then all tangents of the oval meet in a common point, the so-called *nucleus* of the oval. By adding this nucleus to the oval, we arrive at a *hyperoval*, that is, a set of $n + 2$ points such that any line in the plane meets the hyperoval in 0 or 2 points. Conversely, by removing any point from a hyperoval we are left with an oval. There are no hyperovals in projective planes of odd order. Lines that intersect a given (hyper)oval in 0 or 2 points are called *exterior lines* and *secant lines* of the (hyper)oval, respectively. With respect to an oval, the point set of a projective plane of odd order is partitioned into three parts consisting of the *interior points*, the points in the oval, and the *exterior points* of the oval. The interior points are the points that are not contained in any tangent line. Every exterior point is contained in two tangent lines.

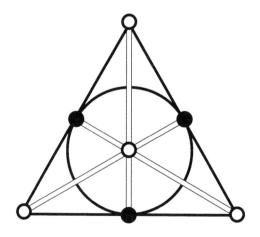

The vertices of the outer triangle in the above picture form an oval in the Fano plane. The lines through the center of the diagram are the tangents of the oval, the center itself the nucleus of the oval. This means that the set consisting of the open points is a hyperoval in our projective plane.

The hyperovals in the Fano plane are the complements of the lines with respect to the point set of the plane. The hyperoval that we just considered, for example, is the complement of the "circle."

Segre's theorem, one of the most influential results in incidence geometry, asserts that the ovals in the classical projective planes of odd order are the nondegenerate conic sections (see [56] for a proof of Segre's theorem). In classical projective planes of even order the situation is much more complicated, and we know of a number of infinite families of ovals that are not conic sections. In fact, one of the most important open problems in incidence geometry is the classification of all ovals in the classical projective planes of even order. Such a classification would have important implications for numerous other kinds of geometries, such as generalized quadrangles, which we are going to consider in Chapter 4.

In Chapter 8 we are going to use ovals and hyperovals to construct pictures of projective planes with the ovals and hyperovals right in the middle of the picture and, in the case of ovals in projective planes of odd order, the sets of interior and exterior points nicely separated out and arranged in a very symmetric fashion.

1.7   Blocking Sets

A *blocking set* in a geometry is a set of points such that every line of the geometry contains at least one point in the set. From this definition it is clear that any set of points containing a blocking set is a blocking set itself. A blocking set is called *minimal* or *irreducible* if it does not contain any smaller blocking set. It is also clear that in a projective plane every line is a (minimal) blocking set. Convince yourself that every minimal blocking set in the Fano plane is a line.

For more information about blocking sets in projective planes in general see [50, Chapter 13] (note that Hirschfeld uses a slightly more restrictive definition of blocking sets). We also give some more examples of blocking sets in the projective plane of order 3 in Section 6.2.1 and in biplanes in Section 10.6.

1.8   Difference Sets and Singer Diagrams

Singer's theorem (see, for example, [56]) states that the classical projective plane of order n admits a fixed-point-free automorphism γ that has only one orbit. This means that the length of this orbit equals the order of γ, which in turn equals $n^2 + n + 1$, the number of points of the plane.

This provides us with the following neat way of drawing a picture based on a regular $(n^2 + n + 1)$-gon having just one generator: Choose one point of the plane and label it 0. For $k = 1, 2, \ldots, n^2 + n$ label the image of the point 0 under the automorphism γ^k by k. Note down the labels of all the points that are contained in the line that contains the points 0 and 1. In the case of the Fano plane one possible set of such labels is $\{0, 1, 3\}$. Draw a regular $(n^2 + n + 1)$-gon and label the vertices by proceeding in a clockwise direction starting with 0. Draw in a curve that contains all the vertices whose labels correspond to those of the line connecting 0 and 1. This is the generator we have been looking for!

Here is one of the possible pictures we arrive at by doing this for the Fano plane. We call a picture like this a *Singer diagram*.

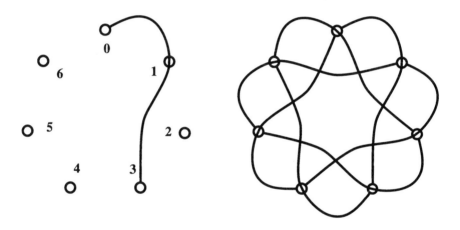

The set of labels corresponding to the line through 0 and 1 is called a *cyclic difference set* of order $n^2 + n + 1$. The name "difference set" alludes to the fact that every positive integer less than $n^2 + n + 1$ can be represented in a unique way as a difference modulo $n^2 + n + 1$ of two different elements in the set. For example, the set $\{0, 1, 3\}$ is a difference set of order 7. We can convince ourselves that this is indeed the case: $1 = 1 - 0$ mod 7, $2 = 3 - 1$ mod 7, $3 = 3 - 0$ mod 7, $4 = 0 - 3$ mod 7, $5 = 1 - 3$ mod 7, $6 = 0 - 1$ mod 7.

A difference set like this contains all the information necessary to reconstruct the whole plane. For example, the lines of the Fano plane are $\{0, 1, 3\}$, $\{1, 2, 4\} = \{0 + 1 \bmod 7, 1 + 1 \bmod 7, 3 + 1 \bmod 7\}$, $\{2, 3, 5\}$, $\{3, 4, 6\}$, $\{4, 5, 0\}$, $\{5, 6, 1\}$, $\{6, 0, 2\}$.

Geometries other than projective planes can also be encoded by difference sets, and we will make use of this fact in later sections.

For more information about difference sets see [3], [45], and [57].

1.9   Incidence Graphs

A difference set corresponding to a classical projective plane of order n can also be used to draw a nice picture of the incidence graph of the plane: Start with a regular $2(n^2+n+1)$-gon and label its vertices in the clockwise direction starting with an open 0, followed by a solid 0, an open 1, a solid 1, an open 2, a solid 2, etc. If $\{0, 1, a, b, c, \ldots\}$ is the difference set, start by connecting the open 1 to the solid 0, 1, a, b, c, etc. The remaining edges of the graph are images of these edges under rotations that rotate open vertices onto open vertices and therefore solid vertices onto solid vertices.

Here is the incidence graph of the Fano plane drawn in this manner.

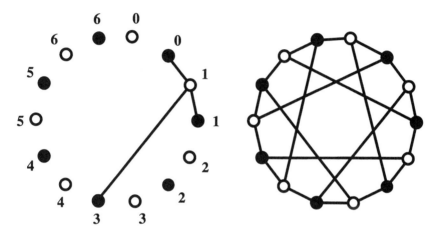

Of course, the vertices of one kind correspond to the points of the projective plane, while the vertices of the other kind correspond to the lines of the projective plane. The reflection through the vertical symmetry axis of the diagram corresponds to a polarity of the projective plane. What are the absolute points and lines of this polarity?

The incidence graphs of projective planes are themselves interesting geometries, namely generalized hexagons (see Chapter 13).

1.10   Spatial Models

Another idea for generating appealing models of geometries is to try to model them on some highly regular spatial objects like the Platonic solids. Here is a first such model of the Fano plane on the tetrahedron.

Remember that a tetrahedron has 4 vertices, 6 edges, and 4 faces. Two edges of the tetrahedron are called *opposite* if they do not have a vertex in common; there are three pairs of such opposite edges.

Take as the points of the geometry the centers of the edges and the center of the tetrahedron. The lines are the circles inscribed in the faces and the line segments connecting opposite edges.

The first of the following pictures shows two generators of our model (the circle and the segment connecting opposite edges of the tetrahedron) with respect to the group of rotations of the tetrahedron. The other two diagrams show the full geometry, inscribed in the tetrahedron and standing by itself.

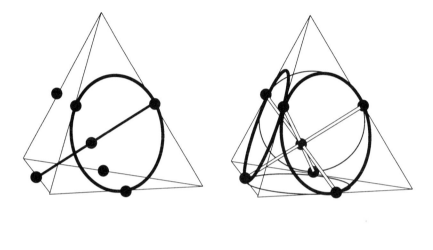

See also in Chapter 14 the photo of this last model built from pipe cleaners.

Our second spatial model is again modelled onto the tetrahedron. This time the points of the plane are the 7 rotation axes of the tetrahedron, or equivalently, 7 pairs of points on the tetrahedron that correspond to the pairs of points of intersections of the 7 axes with the surface of the tetrahedron. The following pictures show the two essentially different lines in this model consisting of three axes each. The first line stands only for itself and the second line for 6 lines in the model.

This last model of the projective plane of order 2 on the tetrahedron has counterparts for the projective planes of orders 3 and 5 on the cube and the dodecahedron, respectively. See [65] for a beautiful exposition of these constructions. We present the models on the cube and the dodecahedron in Sections 6.4 and 8.2.

Note that the four rotation axes that contain vertices of the tetrahedron form a hyperoval in the Fano plane, that the first of the lines below is the only line that does not intersect this hyperoval, and that all the lines of the second kind intersect the hyperoval in 2 points.

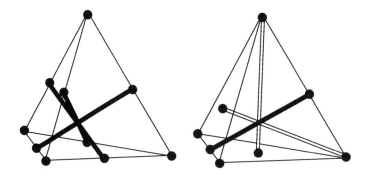

Whenever you try to construct a nice model of a geometry, keep the following important construction principle in mind. It sounds very naive, but it really works more often than not!

Construction principle: number right → everything right

Given a small, highly symmetric geometry with n points, look for the same number of points arranged into a highly symmetric spatial object. Try to merge the two structures such that the symmetries of the spatial object translate into symmetries of the geometry.

2
Designs

We call a geometry a $t - (v, k, \lambda)$ design if it contains v points, if every line, or *block*, contains k points, and if given any t distinct points, there are exactly λ blocks containing them. A $2 - (v, k, \lambda)$ design is called *symmetric* if the number of blocks equals the number of points in the design. The dual of a symmetric design is also a symmetric design with the same parameters.

The projective planes of order n are exactly the $2 - (n^2 + n + 1, n + 1, 1)$ designs, and the affine planes of order n are exactly the $2 - (n^2, n, 1)$ designs. The finite projective planes are examples of symmetric designs. This means that there are infinitely many symmetric $2 - (v, k, 1)$ designs. On the other hand, only finitely many symmetric $2 - (v, k, \lambda)$ designs are known to exist for any given value of $\lambda > 1$.

The symmetric $2 - (v, k, 2)$ designs are straightforward generalizations of the projective planes. They are called *biplanes*. We dedicate the whole of Chapter 10 to pictures of biplanes.

The *unitals* are the $2 - (n^3 + 1, n + 1, 1)$ designs for $n > 1$. For $n = 2$, there is a unique $2 - (9, 3, 1)$ design. This is just the affine plane of order 3 (see Section 7.2).

A construction of symmetric $2 - (q^3 + q^2 + q + 1, q^2 + q + 1, q + 1)$ and $2 - (q^2(q + 2), q(q + 1), q)$ designs from generalized quadrangles is described in Section 4.6.

The *inversive planes* of order q are just the $3 - (q^2 + 1, q + 1, 1)$ designs. We come across the inversive planes of orders 2 and 3 in Chapters 5 and 12.

As t gets larger it becomes more and more difficult to find t-designs. A couple of highly symmetric 5-designs have been known for a long time, but only recently have t-designs for $t > 5$ been found. We describe the construction of the classical $5-(12,6,1)$ design as an extension of the affine plane of order 3 in Section 6.5 and say something about the connection between the classical $5-(24,8,1)$ design and the projective plane of order 4 at the beginning of Chapter 7.

The derived geometry of a $t-(v,k,\lambda)$ design at a point is a $(t-1)-(v-1,k-1,\lambda)$ design.

For some excellent introductions to designs see [7], [17], [23], [24], and [57]. See also the "Handbook of combinatorial designs" [25] for survey articles and comprehensive lists of designs.

2.1   The Smallest Nontrivial 2-Design

The smallest nontrivial 2-design is a $2-(6,3,2)$ design. This design has 6 points and 10 blocks. Cameron [23] gives the following description: The six points are the center and vertices of a pentagon. The blocks consist of all triangles formed from these points that contain exactly one edge of the pentagon. This description immediately translates into the following generator-only picture for this design.

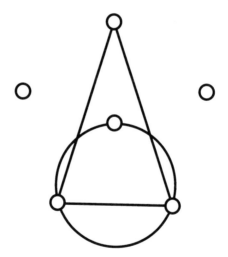

It is also possible to model this design onto the icosahedron: The points of the design are the six axes connecting opposite vertices of the icosahedron (remember that an icosahedron has 12 vertices). To every face of the icosahedron and its opposite there corresponds a block consisting of the three axes that contain the vertices of the face (and its opposite). Since there are 20 faces, we really end up with 10 blocks.

2.2 Hadamard Designs

Apart from being the smallest projective plane, the Fano plane is also the smallest example of a *Hadamard 2-design*, that is, a $2 - (4\lambda + 3, 2\lambda + 1, \lambda)$ design. Hadamard designs are quite easy to construct. Still, it is not known whether they exist for all λ. For $\lambda = 2$, there is a unique Hadamard 2-design with parameters $2 - (11, 5, 2)$. This is a biplane we have a closer look at in Section 10.2. The geometry of subplanes of the three-dimensional projective space of order 2 is an example of a $2 - (15, 7, 3)$ design which is a Hadamard 2-design for $\lambda = 3$. We come across this geometry in Chapter 5. More generally, the geometry of hyperplanes of the finite n-dimensional projective space of order 2 is a Hadamard 2-design for $\lambda = 2^{n-1} - 1$.

Given a Hadamard $2 - (4\lambda + 3, 2\lambda + 1, \lambda)$ design, it is possible to construct a *Hadamard 3-design*, that is, a $3 - (4\lambda + 4, 2\lambda + 2, \lambda)$ design as follows. Take as the points of the new design the points of the 2-design plus one additional point. The blocks of the new design are the blocks of the 2-design that have all been extended by the additional point plus all the complements of the blocks of the 2-design (with respect to the whole point set of the 2-design). We note that every Hadamard 3-design arises from a Hadamard 2-design in this way. The Hadamard 3-design constructed from a Hadamard 2-design is also called a *one-point extension*. The derived geometry of a Hadamard 3-design at any of its points is a Hadamard 2-design.

2.2.1   The One-Point Extension of the Fano Plane

Here we give a pictorial representation of the one-point extension of the Fano plane. All the derived geometries of this $3 - (8, 4, 1)$ design are indeed Fano planes, as we shall see.

Consider the following geometry. Its points are the 8 vertices of a cube. It has 14 blocks, containing 4 points each: the vertex sets of the 2 tetrahedrons as shown in the first of the following diagrams, the 6 vertex sets of the faces of the cube, and the 6 vertex sets of the "diagonal rectangles" as shown in the last of the following diagrams.

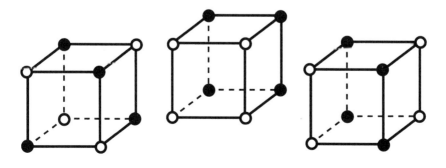

It is clear that the automorphism group of this design acts transitively on the point set. We show that the derived geometry at any point is indeed the Fano plane. For this we look at the cube in the direction of one of the main diagonals of the cube. What we see is the following.

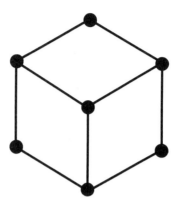

The following diagram on the left is the generator-only diagram of the derived geometry at the point of the cube that we do not see. We deform this model by moving three points as indicated by the arrows to arrive at the traditional generator-only model of the Fano plane.

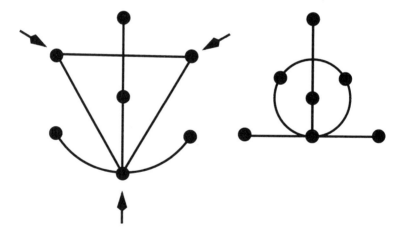

If you are familiar with the representation of the classical projective planes and spaces via homogeneous coordinates, it should be clear how to describe this model in a simple algebraic fashion that immediately suggests how to represent the geometry of hyperplanes of higher-dimensional projective spaces of order 2: In homogeneous coordinates the points of the Fano plane are just the nonzero vectors of length three over $\mathbf{Z}_2$. These vectors correspond in a natural way to the points of the cube as follows. The point $(0,0,0)$ will be the point that gets added to the points of the 2-design when we construct the 3-design.

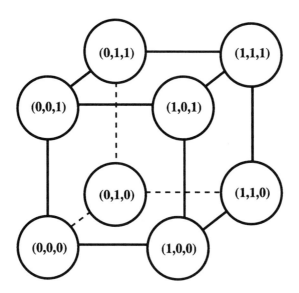

Now, to every point $(a, b, c) \neq (0, 0, 0)$ there correspond two blocks of the 3-design. The first one is the set of all points (x, y, z) such that $ax+by+cy = 0 \bmod 2$. The second one is its complement. Verify that this construction yields the above one-point extension of the Fano plane.

2.3   Steiner Triple Systems

A $2 - (v, 3, 1)$ design is also called *Steiner triple system*, or for short STS. The number v is the *order* of the STS. It turns out that an STS of order v exists if and only if $v = 1$ or $3 \bmod 6$. If we identify the v points of an STS with the vertices of the complete graph with v vertices, then every block of the STS corresponds to a triangle of edges of the graph, and the STS itself can therefore be regarded as a partition of the edges of this complete graph into triangles. For a very good introduction to STSs see [23].

The Fano plane is the (unique) smallest nontrivial STS. It is of order 7. There is also just one STS of order 9, namely the affine plane of order 3 that we are going to consider in Chapter 6. Here are two generator-only pictures of these two STSs.

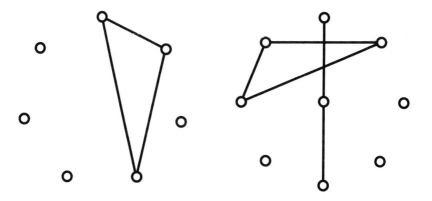

There are two essentially different STSs of order 13. Here is a generator-only model of one of them.

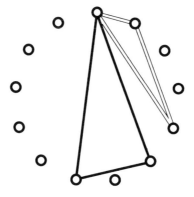

The second one can be constructed by replacing in this STS the following 4 blocks on the left by the 4 triangles on the right.

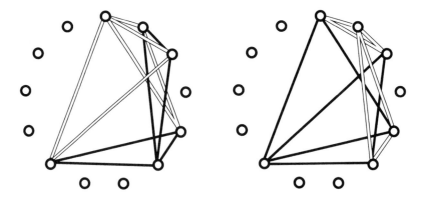

Here is a generator-only picture of one of the 80 different STSs of order 15.

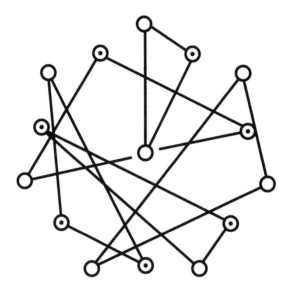

This STS of order 15 is the geometry of points and lines of the three-dimensional projective space over the field with two elements that we are going to consider in some more detail in Chapter 5. In general, the geometry of points and lines of an n-dimensional projective space over the field with two elements is an STS of order $2^{n+1} - 1$, and the n-dimensional affine space over the field with three elements is an STS of order 3^n.

2.3.1 Kirkman's Schoolgirl Problem

Our generator-only picture of the STS of order 15 also contains a solution of the (classical) *Kirkman schoolgirl problem*:

Fifteen schoolgirls walk each day in five groups of three. Arrange the girls' walks for a week so that in that time, each pair of girls walks together in a group just once.

Note also that every one of the 15 points of this STS is contained in exactly one of the generators in the picture. By letting the points of the STS correspond to the 15 girls and the 7 images under rotation of the set of generators to the walks, we get a solution to our problem. We remark that this solution to our problem is by no means unique.

We generalize the problem by replacing "fifteen" by a number $3n$ and "five" by n. It can be shown that this generalized Kirkman problem has a solution. In the case $n = 3$ the (unique) solution corresponds to the set of parallel classes in the affine plane of order 3. One such parallel class is given in the following diagram.

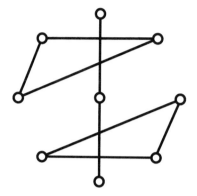

3
Configurations

An *abstract plane configuration*, for short APC, with parameters (p_n, l_m) is a geometry that satisfies the following axioms:

Axioms for abstract plane configurations

(C1) There are p points and l lines with n lines through every point and m points on every line.

(C2) Two distinct points are contained in at most one line.

(C3) Two distinct lines intersect in at most one point.

(C4) The geometry is connected.

An easy counting argument shows that in an abstract plane configuration like this $pn = lm$. If such a geometry can be represented in the Euclidean plane with straight line segments as lines, we omit the word "abstract," or the "A" in the abbreviation. If the number of points equals the number of lines and the number of points per line the number of lines through every point, then we abbreviate (p_n, l_m) by (p_n).

The Fano plane is an APC with parameters (7_3) that is not a PC. More generally, a projective plane of order n is an APC with parameters $(n^2 + n + 1_{n+1})$ that is not a PC. Similarly, an affine plane of order n is an APC with parameters $(n^2{}_{n+1}, n^2 + n_n)$ that is a PC only if $n = 2$. The last two facts follow from *Sylvester's theorem*:

It is not possible to arrange any finite number of real points so that a straight line through every two of them shall pass through a third, unless they all lie in the same straight line.

See [27, Section 4.7] for some historical information and a short proof of this result. One of the best introductions to configurations is the respective chapter in [49].

Here is an example of an APC with parameters $(16_3, 12_4)$.

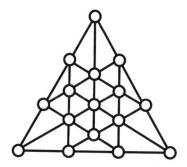

Let us have a look at the APCs with small parameters (p_n). If $n = 2$, then for every p greater than three, there is exactly one such APC, namely the p-gon.

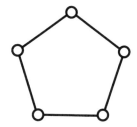

3.1   Configurations with Three Points on a Line

The case $n = 3$ is much more interesting. The following table lists the numbers of APCs with parameters (p_3) for small p's (see [8]).

p	7	8	9	10	11	12	13	14
no.	1	1	3	10	31	229	2036	21399

We are going to present pictures for all such APCs on up to 10 points. We remark only that all APCs with parameters 11_3 and 12_3 are PCs (see [112]).

3.1.1   The Fano, Pappus, and Desargues Configurations

By far the most important APCs are the Fano plane, the Pappus configuration, and the Desargues configuration.

The Pappus configuration is a PC with parameters (9_3) and is usually depicted as follows.

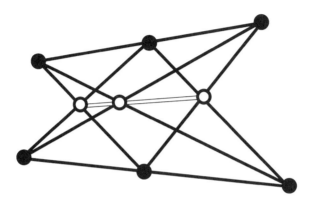

The Desargues configuration is a PC with parameters (10_3) and is usually depicted as follows.

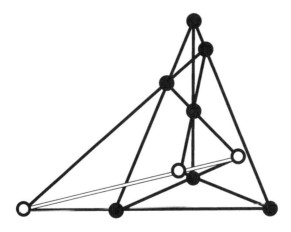

Remember that classical projective planes are characterized by the fact that all possible Desargues configurations "close." Because of this fact the classical affine and projective planes are also called Desarguesian. For the Desargues configuration to close means just that if one draws the solid black parts of the configuration, then the three open points of intersection are collinear. Similarly, all Pappus configurations close in all classical projective planes, and all Fano configurations close in classical projective planes of even order. As with the Desargues configuration, to close means just that if one draws the solid black parts of such a configuration, then the three open points of intersection are collinear.

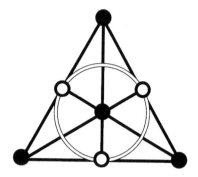

3.1.2   The Configurations with Parameters (7_3) and (8_3)

What other APCs with parameters (p_3) are there? It is clear that p has to be at least 7. It turns out that the only APC with parameters (7_3) is the Fano plane (see again the following generator-only picture on the left). There is also a unique APC with parameters (8_3). We can construct it by removing a point together with all the lines through this point from the affine plane of order 3. If we choose this point to be the center of the model for this plane on page 24, then we arrive at the following generator-only model on the right.

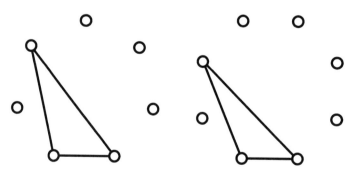

The APC with parameters (8_3) is not a PC. The best picture of this geometry, using as many straight line segments as possible as lines of the geometry, looks as follows.

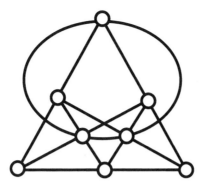

3.1.3   The Configurations with Parameters (9_3)

Apart from the Pappus configuration, there are two further APCs with parameters (9_3). The generator-only models of the Fano plane and of the APC with parameters (8_3) suggest the following way to construct a generator-only model of an APC with parameters (p_3) for any p greater than 6: Take as the points the vertices of a regular p-gon and choose as the generator a triangle as in the two generator-only models above. If we do this for $p = 9$, we arrive at a configuration with the same parameters as the Pappus configuration. In the picture in the middle we connect two points of the configuration if and only if they are not contained in a line of the geometry. This gives a 9-gon. The picture on the right shows a representation of this particular configuration using straight-line segments. This means that we are dealing with a PC.

We represent the Pappus configuration on the regular 9-gon. The 9 triangles in the small 9-gons on the left are the lines. In the picture in the middle we connect two points of the configuration, as above, if and only if they are not contained in a line of the configuration. This gives three triangles, which implies that this configuration is really different from the one that we considered before. The picture on the right is again the traditional picture of the Pappus configuration.

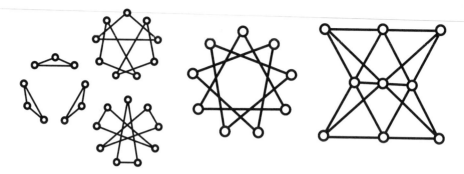

Finally, here is the corresponding sequence of pictures for the last of the three APCs with parameters (9_3).

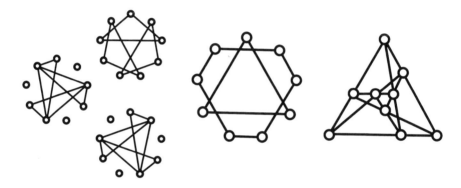

Assuming that there are really only three different APCs with parameters (9_3), convince yourself that the straight-line models on the right are really isomorphic to the respective circular models on the left: Just connect pairs of noncollinear points by an edge and see whether you arrive at a 9-gon, a hexagon, and a triangle, or three triangles.

3.1.4   The Configurations with Parameters (10_3)

Apart from the Desargues configuration there are 9 further APCs with parameters (10_3), 8 of which are PCs. We first present two pictures each for every single one of the 8 PCs. The pictures on the right are the straight-line pictures of the different configurations in [13]. The corresponding pictures on the left are highly symmetric representations that usually exhibit most of the symmetries of the configurations. These pictures can be found in [9].

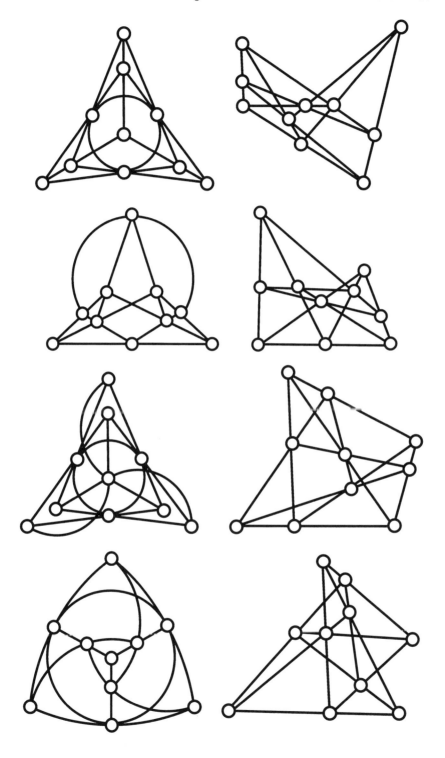

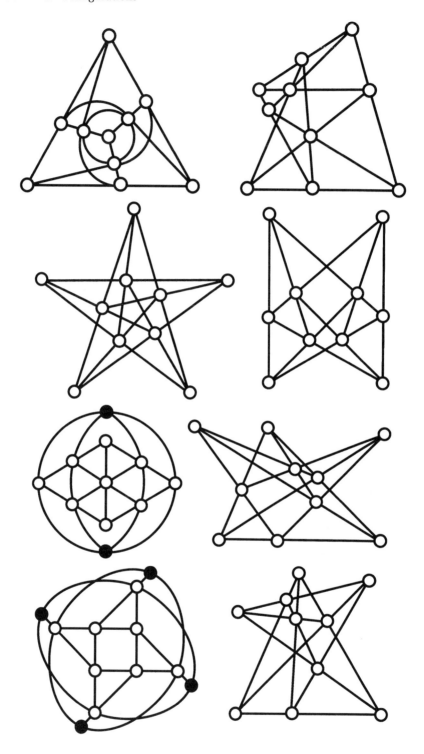

In the last two pictures, pairs of "antipodal" solid points are supposed to be identified.

The following picture shows the only APC with parameters (10_3) that is not a PC.

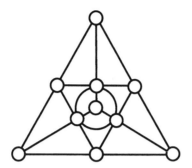

This last APC also has a very appealing representation on the tetrahedron: The points of the geometry are the vertices and the centers of the edges of the tetrahedron. With every vertex of the tetrahedron we associate a line consisting of the centers of the three edges containing the vertex. The edges of the tetrahedron are the remaining 6 lines of the APC.

Note that the 6 centers of the edges of the tetrahedron are the vertices of an octahedron. The faces of this octahedron can be coloured, 4 in black and 4 in white, such that adjacent faces have different colours and such that the 4 vertex sets of the black faces are the 4 distinguished lines in the above APC model. Furthermore, the vertex sets of both the black and the white faces are APCs with parameters $(6_2, 4_3)$, and the tetrahedron itself is an APC with parameters $(4_3, 6_2)$. There is just one APC with parameters $(6_2, 4_3)$. It is called the *complete quadrangle*. The dual of the complete quadrangle is the unique APC with parameters $(4_3, 6_2)$. This means that the above APC is the union of a complete quadrangle and its dual. The Desargues configuration can also be seen to be such a union.

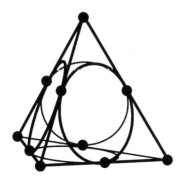

See page 124 for stereograms of this model and the related model of the APC described above.

3.2 Configurations with Four Points on a Line

The following table lists the numbers of APCs with parameters (p_4) for small p's. A summary of representations of these APCs can be found in [8].

p	13	14	15	16	17
no.	1	1	4	19	1972

The APC with parameters (13_4) is the projective plane of order 3. We are going to present a number of pictures for this projective plane in Chapter 6.

The projective plane of order 4 is made up of 21 points and contains lots of Fano configurations. Let us concentrate on one such Fano configuration. The APC with parameters (14_4) consists of the 14 points that are not points of the configuration and the 14 lines that intersect the configuration in one point each. The authors of [8] attribute this neat construction to D. Glynn. We present pictures of partitions of the projective plane of order 4 into 3 Fano planes in Chapter 7. Have a close look at these pictures and see whether you can identify our configuration.

Here are three generator-only models for different APCs with parameters (15_4).

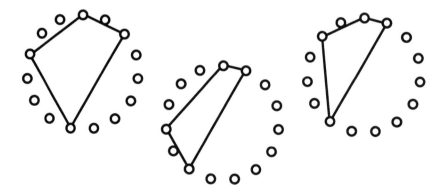

Let l be a line in the projective plane of order 4 and let p be a point not contained in l. If we delete p, l, all lines through p, and all points of l from the projective plane, we are left with the APC with parameters (15_4) on the far right above.

The remaining APC with these parameters occurs as a substructure of the projective plane of order 5 (see [8] for details). The last two constructions are again due to D. Glynn.

3.3   Tree-Planting Puzzles

Here is a famous puzzle that is attributed to Isaac Newton (see [32, p. 56] and [35, p. 18]): Plant nine trees so that they shall form ten straight rows with three trees in every row. If we were looking only for nine straight rows, then any of the PCs with parameters (9_3) would give a solution to this problem. It turns out that to the traditional picture of the Pappus configuration in its most symmetric form, one more straight line segment can be added that contains three points of the configurations. The resulting picture is a solution to the puzzle.

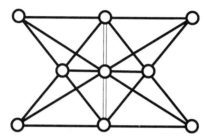

Of course, the geometry that corresponds to this picture is no longer a configuration. Still, this and solutions to other tree-planting puzzles can be used to construct beautiful pictures of some of the other geometries we are interested in. For example, by adding two lines (no longer straight) to the above picture, we arrive at the following picture of the affine plane of order 3.

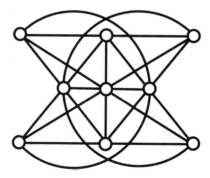

Can you see the parallel classes in this affine plane?

Is there a way to "measure" how beautiful a picture of a given geometry is? One obvious way of doing this would be to count the number of straight-line segments in the picture that actually correspond to lines of the geometry. If we make up our minds that this is what we are after, then the above picture is probably the most beautiful picture of the affine plane of order 3. The most naive way of trying to find the most beautiful picture of a "linear" geometry with p points and n points on every line is to try to plant p trees such that a maximum number of rows with n trees in each row occurs. Requiring that there be a maximum number of such rows seems to guarantee that the corresponding diagram will be highly symmetric. Now, with a little bit of luck we will be able to complement this picture to a picture of the geometry we are interested in by adding some skew lines. Of course, there are a number of problems with this approach. For example, how do we tell whether a given arrangement of points is maximal? Anyway, here are two arrangements of points that seem to be maximal for the respective parameters.

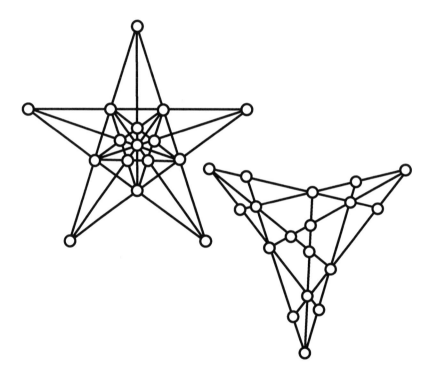

In the PC on the left $p = 16$ and $n = 4$. It looks very much as if the affine plane of order 4 can be built around this picture. We will show that this is really possible in Chapter 9. On the other hand, the PC with parameters $(16_3, 12_4)$ at the beginning of this section cannot be extended to an affine plane of order 4. Why not? We do not know what the PC on the right is good for. Can you think of anything?

For more tree-planting puzzles see [16].

4
Generalized Quadrangles

A geometry is a *generalized quadrangle* if it satisfies the following axioms:

Axioms for generalized quadrangles

(Q1) Two distinct points are contained in at most one line.

(Q2) Given a line l and a point p not on l, there is exactly one line k through p that intersects l (in some point q).

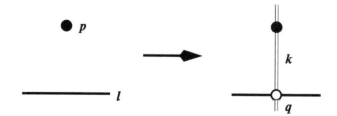

A finite generalized quadrangle is of *order* (s, t) if every line contains $s+1$ points and every point is contained in exactly $t + 1$ lines. A geometry like this has $(s + 1)(st + 1)$ points and $(t + 1)(st + 1)$ lines. The name "generalized quadrangle" has to do with the fact that an ordinary quadrangle is a generalized quadrangle and that, just as in this prototype, it is not possible to draw any digons or triangles in a generalized quadrangle by just using lines of the geometry. It is possible to draw quadrangles.

The regular quadrangle is a generalized quadrangle of order $(1,1)$. The generalized quadrangle of order $(n,1)$ is a square grid.

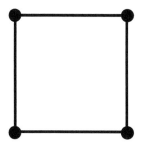
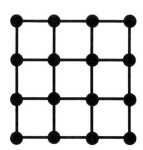

The dual of a generalized quadrangle of order (s,t) is a generalized quadrangle of order (t,s). The generalized quadrangle of order $(1,n)$ is the *complete bipartite graph* on $2n$ vertices. This graph can be described as follows. Its vertices are the vertices of a regular $2n$-gon that have been labelled (in a natural way) from 0 to $2n-1$. Two vertices i and j are connected by an edge if and only if $i-j$ is an odd number. The following two pictures show the generalized quadrangles of orders $(1,3)$ and $(1,4)$.

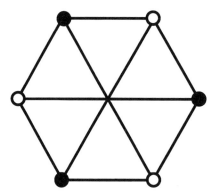
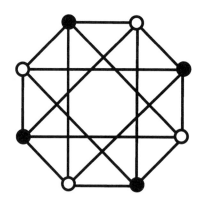

The generalized quadrangle of order $(1,4)$ can also be modelled on the cube as follows: Take as its points the 8 vertices of the cube and as its lines the 12 sides and 4 main diagonals of the cube.

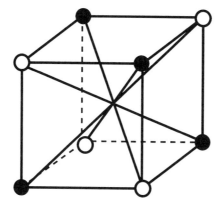

4.1   The Generalized Quadrangle of Order $(2,2)$

The generalized quadrangle of order $(2,2)$ is the smallest nontrivial example of a generalized quadrangle. It is uniquely determined and, just like the different models of the Fano plane, a number of models of this geometry play an important role in the following.

4.1.1   A Plane Model—The Doily

Here is a well-known model of this generalized quadrangle that was invented by S. Payne, one of the world's leading experts on generalized quadrangles. Its first appearance in print was on the cover of the proceedings of a conference that was held in 1972 (see [73]). Payne calls this particular picture the "doily."

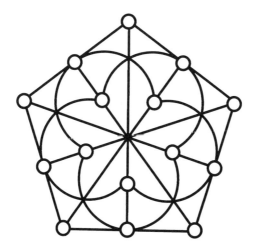

The following three diagrams show the different line pencils in this geometry. With these diagrams it is easy to see that Axiom Q1 is satisfied. In order to check that the second axiom is satisfied, convince yourself that in any single one of the three diagrams a solid black line, that is, a line that does not contain the distinguished point p, intersects exactly one of the lines in the line pencil.

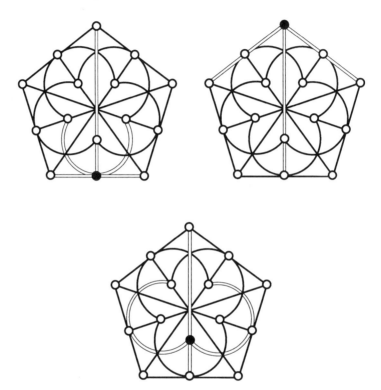

Polarities, Ovoids, and Spreads

You probably know this puzzle: What is the maximum number of queens that can be placed on a chessboard such that no queen attacks another? (see [32, Problem 300]). The answer is 8, and there are 96 such arrangements. Now, instead of placing the queens on a chessboard, we put them on the doily. We say that two queens attack each other if they are standing on points that are collinear. Now, what is the maximum number of queens that can be placed on the doily such that no queen attacks another, and how many different such arrangements are there? By the end of this section you should know the answer to this question.

The following diagram shows three generator-only models that describe a polarity of our geometry. Each diagram highlights a point and a line that get exchanged by the polarity.

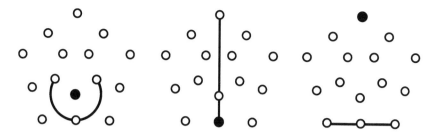

The absolute points of this polarity, that is, the points that are contained in their images, form an ovoid, that is, a set of points that has exactly one point in common with every line. The absolute lines of the polarity form a *spread*, that is, they partition the point set. See the following diagram on the left. Apart from the above polarity, our geometry admits a further 5 polarities (see [74, p. 122]). The diagram on the right shows what the corresponding ovoids and spreads look like.

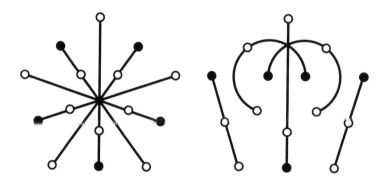

Note that any two of the 6 ovoids intersect in exactly one point. This will prove important when we extend the plane model of the generalized quadrangle to a model of the projective plane of order 4 in Chapter 7 by using these ovoids as lines of the projective plane.

Note also that the line pencils through points of an ovoid partition the line set of the generalized quadrangle; that is, every line in our geometry is contained in exactly one of these line pencils.

4.1.2 A Model on the Tetrahedron

We present a three-dimensional model of our generalized quadrangle on the tetrahedron. Its points are the center and the vertices of the tetrahedron plus the centers of its edges and faces. The lines are the medians of the faces plus the three segments connecting the centers of opposite edges of the tetrahedron.

Note that in the following diagram the edges of the tetrahedron are drawn in only for reference.

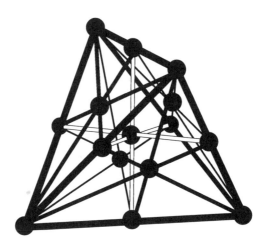

Ovoids and Spreads

The 6 ovoids of the generalized quadrangle can also be easily described in this model:

- The 2 vertex sets of tetrahedrons together with the center of the model. The first such tetrahedron is the tetrahedron of reference (see the following picture on the left). The second one is the tetrahedron formed by the centers of the faces of the tetrahedron of reference.

- The 4 "spikes." The following figure on the right shows one of them. The other 3 can be constructed by rotating the tetrahedron of reference.

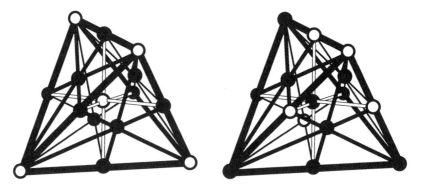

Modulo the rotations of the tetrahedron of reference, all spreads of the generalized quadrangle look as follows.

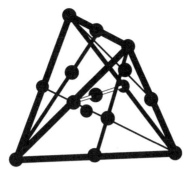

Stereogram

In this section we present a stereogram picture pair of our generalized quadrangle created by Andreas Schroth. There are basically three different techniques to view stereograms. Note that any given stereogram has been built with a special viewing technique in mind and can usually only be viewed with this particular technique.

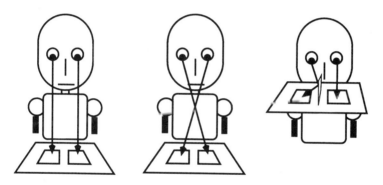

The first one is called the *parallel technique*: For this your right eye focuses on the right picture and your left eye focuses on the left picture. This is achieved by staring "through" the pictures at a point at infinity.

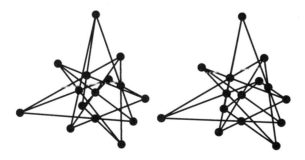

With the *cross-eyed technique* look at the left picture with your right eye and at the right picture with your left eye. To achieve this, hold a pencil between your eyes and the book. Focus on the tip of the pencil. While keeping your eyes fixed on its tip, slowly move the pencil towards the book. Once the correct position is reached, the image should come into view.

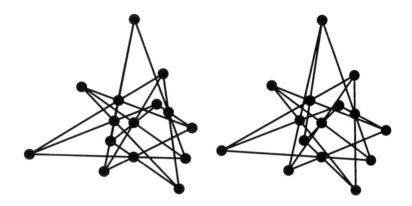

With the *mirror technique* hold a mirror between the two pictures. You look at the right picture in the mirror and at the left picture on the page. This is probably the easiest method of viewing stereograms.

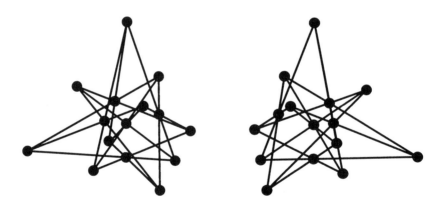

Apart from the three stereograms above, all stereograms in the main part of this book can be viewed with the parallel as well as the cross-eyed technique. If you have mastered both techniques, you will notice that the three-dimensional pictures get turned "inside out" if you switch from one technique to the other. If you are shortsighted and prefer the parallel technique, try viewing the stereograms without your glasses on. While most people have no problems viewing stereograms with either the parallel or the cross-eyed technique, some people, especially those with eye problems other than short- or farsightedness, find it impossible to make any stereogram work. If you have serious problems, do not give up before you have tried the mirror technique described above. If this works for you, have a look at Appendix B. There all stereograms that you come across in the main part of the book are listed again suitable for viewing with the mirror technique. As with switching from the parallel to the cross-eyed technique, flipping the mirror over to the other side turns the stereograms inside out.

At the turn of the century stereograms were very much in fashion. Ask around in your local antique shops whether they have any stereogram viewers from that period. All stereograms that can be viewed with the parallel technique can be looked at with these viewers, and all people I know who have never been able to view stereograms have had no problems viewing them with a device like this.

Further information on viewing stereograms can be found in the following beautiful books on stereograms: [54], [55].

A Fake Generalized Quadrangle

In this section we present an appealing spatial model of an abstract plane configuration with the same parameters as our generalized quadrangle, namely (15_3).

The points of the geometry are the center, the centers of the faces, and the vertices of the cube. Its lines are the diagonals of the faces and the segments connecting centers of opposite faces. It is easy to spot lots of triangles in this geometry.

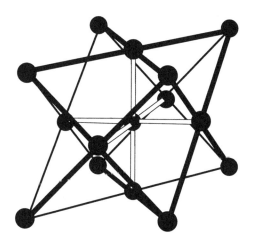

Since it contains triangles, we can be sure that this geometry is not our generalized quadrangle in disguise.

4.1.3   A Model on the Icosahedron

The following abstract description of our generalized quadrangle can be found in [74, p. 122]:

Sylvester's syntheme–duad geometry

A *duad* is an unordered pair $ij = ji$ with $i, j \in \{1, 2, 3, 4, 5, 6\}$ and $i \neq j$. A *syntheme* is a set $\{ij, kl, mn\}$ of three duads for which $i, j, k, l, m,$ and n are distinct. *Sylvester's syntheme–duad geometry* with duads as points and synthemes as lines is isomorphic to the generalized quadrangle of order $(2, 2)$. Equivalently, the points are the transpositions and the lines are the remaining odd involutions in the symmetric group on 6 elements.

The magic number 6 in our abstract description occurs in the icosahedron in the form of the 6 axes connecting opposite vertices. Take any pair, or "duad," of such axes. Then there is exactly one pair of opposite edges of the icosahedron connecting the vertices contained in the axes.

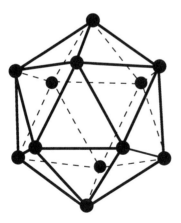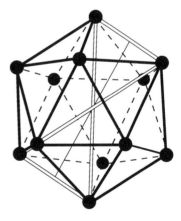

Therefore, it is natural to consider the 15 pairs of opposite edges, or, equivalently, the 15 rotation axes through the centers of the edges, as the points of our quadrangle. The following lines in the quadrangle stand for 5 and 10 lines in the quadrangle, respectively.

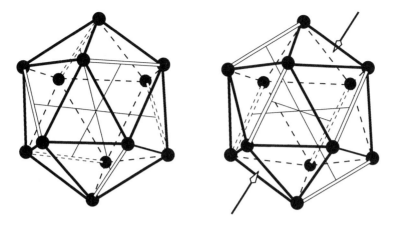

Note that the axes in the first kind of line are mutually perpendicular. In the picture of the second kind of line we have also highlighted an axis through the center of two faces (the two ends of this axis are sticking out at the top and at the bottom of the picture). This axis is not one of the points in the line, but it is distinguished with respect to the three axes in the line in that rotation through 120 degrees around this axis leaves the line invariant. There are 20 faces corresponding to 10 rotation axes, which in turn correspond to the ten lines of the second kind.

Ovoids and Spreads

We note that in the abstract model every integer i from 1 to 6 is associated with an ovoid that consists of all duads of the form ij, $i \neq j$. Therefore, an ovoid in our model corresponds to the set of five edges meeting in a vertex of the icosahedron (see the figure on the left). A spread corresponds to a colouring of the icosahedron with five "colours" as in the figure on the right.

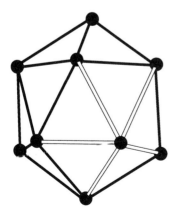
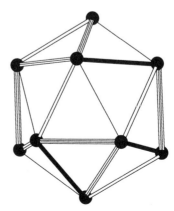

4.1.4   A Model in Four-Space

We start with six points in general position in real projective four-space, say
1, 2, 3, 4, 5, and 6. They determine the 15 points of the generalized quadrangle such as p_{12}, the intersection of the line 12 with the hyperplane 3456.
The lines of the space that contain 3 of these 15 points are the lines of the
quadrangle (see [26]). Our model on the tetrahedron arises as a projection
of this four-dimensional model into real three-space.

4.2   The Petersen Graph

The following diagram shows the so-called *Petersen graph.*

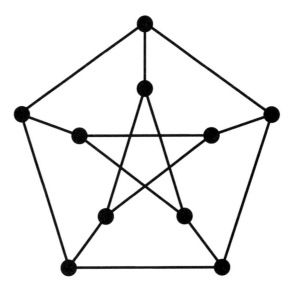

It is one of the most important graphs, and recently a whole book has
been dedicated to it (see [53]). Here are some facts that make it important
for geometers.

- It is the graph of the Desargues configuration. This means that given
  a picture of the Desargues configuration, a picture of the Petersen
  graph can be constructed as follows: Take as the vertices of the graph
  the 10 points of the configuration and connect two vertices by an edge
  if and only if they are not collinear. Actually, according to [53, p. 1],
  the Petersen graph made its first appearance in the literature in just
  this guise (see [62]).

- It is the unique $(3,5)$-cage. See Section 13.6 for more information about this aspect.

- The dodecahedron, viewed as a graph, is the double cover of the Petersen graph. Equivalently, the Petersen graph can be modelled onto the dodecahedron as follows: The vertices of the graph are the 10 axes connecting opposite vertices of the dodecahedron (remember that a dodecahedron has 20 vertices). To every edge of the dodecahedron and its opposite there corresponds an edge of the Petersen graph consisting of the two axes that contain the vertices contained in the edge (and its opposite). Since the dodecahedron has 30 edges, we really end up with 15 edges of the graph.

- If we remove the points of an ovoid in the generalized quadrangle of order $(2,2)$, we are left with the Petersen graph.

Using the last remark, we can rebuild the generalized quadrangle of order $(2,2)$ around the above picture of the Petersen graph. A. Pasini showed me this picture:

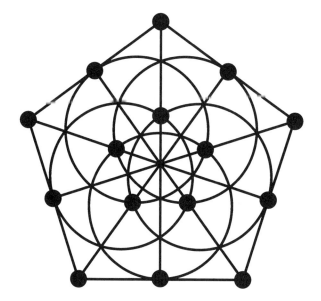

On the other hand, by using the diagrams on pages 41 and 44, we arrive at some two- and three-dimensional pictures of the graph. The two-dimensional ones do not look too unlike the traditional picture of the Petersen graph on the pentagon, but one of the three-dimensional pictures is definitely worth drawing. We construct it by removing the ovoid consisting of the center and the vertices of the tetrahedron. What we are left with is the Petersen graph modelled onto an octahedron.

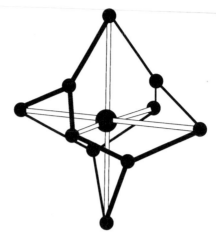

The vertices of the graph are the 6 vertices of the octahedron plus the centers of four of its faces. Note that the center of the model is not a vertex of the graph. There are two kinds of edges:

- A segment connecting a vertex of the graph, which is the center of a face, to a vertex of the same face. There are 12 edges like this.

- A diagonal of the octahedron. There are 3 edges like this.

We describe a connection between the smallest 2-design, that is, the $2 - (6, 3, 2)$ design we already encountered in Section 2.1 and the Petersen graph. Here again is the generator-only picture for this design.

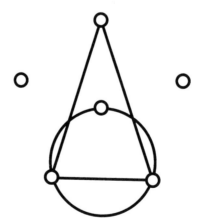

Define a graph whose ten vertices are the ten blocks of the design. Two vertices/blocks are adjacent if and only if they intersect in exactly 2 points. As we can see, this graph is the Petersen graph.

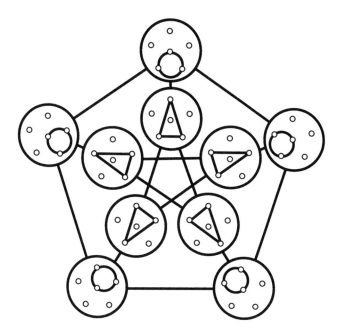

4.3   How to Construct the Models

The way we constructed the model on the icosahedron was well motivated. On the other hand, how did Payne construct his "doily" and how did we come up with the model of the tetrahedron? We now describe a nice way of constructing these symmetric models for the generalized quadrangle of order (2,2) that is again based on its representation in terms of duads and synthemes. Remember that the generalized quadrangle is self-dual. This means that it is possible to use the synthemes as points of the geometry and the duads as lines, where two points/synthemes are connected by a line/duad if and only if the duad in contained in both synthemes. Consider the complete graph on 6 vertices. Then the duads can be regarded as the edges of this graph and synthemes as the spreads, or *1-factorizations*, of this graph. There are different ways in which to arrange 6 points in the plane and in space. Here are some highly symmetric ones: the vertices of a regular pentagon together with its center, the vertices of a regular hexagon, the vertices of an octahedron, the vertices of a prism with a regular triangle as a base.

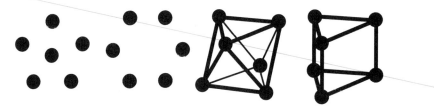

Let us concentrate on the pentagon. Then, up to rotation, we arrive at the following three 1-factorizations of the complete graph, each one of which represents five different factorizations.

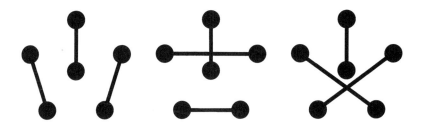

Starting with these factorizations, it is very easy to come up with the traditional plane model of the generalized quadrangle we have been working with so far.

Now we concentrate on the hexagon. Here we get the following five different factorizations which represent 1, 2, 6, 3, and 3 factorizations, respectively.

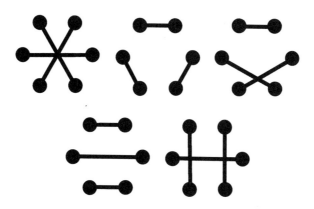

We can label our three-dimensional model using these factorizations as labels to see the automorphism of order 3 of the generalized quadrangle reflected in these factorizations. For this we look at the model from "above." What we see is the following. Note that two points get obscured by the point in the middle of the diagram.

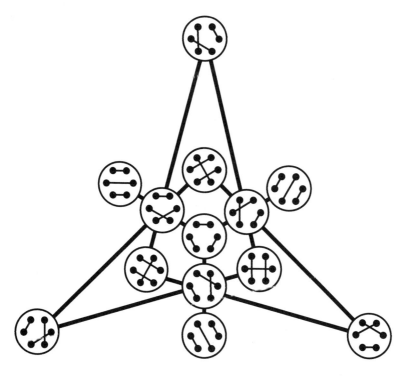

What we do not see are the line pencils through the two obscured points. Here are the two missing pencils.

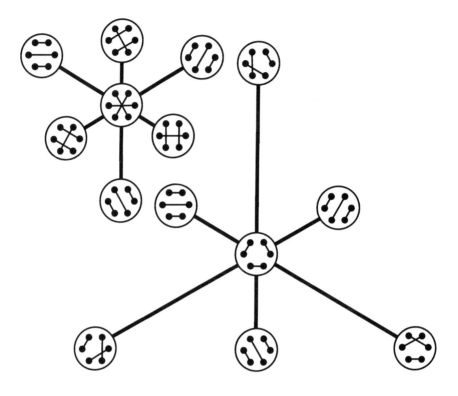

Construct the different kinds of factorizations for the octahedron and the prism and see whether you can use them to construct new models of the generalized quadrangle or whether you can label one of our old models in new, meaningful ways.

Finally, note that the union of the edges in two factorizations that do not have an edge in common always forms a cycle of length six. For example:

We can use the same kind of argument to derive meaningful labellings of the different models of the Petersen graph, since this graph is just the "geometry of duads and synthemes on 5 labels." We start by listing four symmetric arrangements of 5 points: the vertices of the regular pentagon, a square together with its center, a double pyramid with a regular triangle as base, the vertices of the 5-cell.

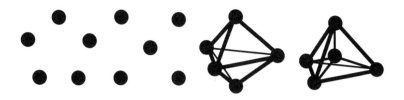

Let us concentrate on the regular pentagon. The following two edges stand for essentially different kinds of edges in the complete graph on five vertices.

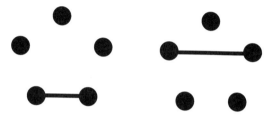

Here is a meaningful labelling of our favourite model of the Petersen graph using the different edges of the complete graph on the vertices of the pentagon as labels. Note that two vertices of the graph are adjacent if and only if the edges in their labels have no points of the pentagon in common.

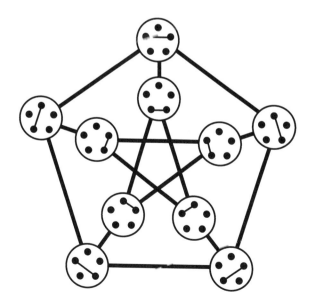

Try to construct meaningful labellings of our three-dimensional model of the Petersen graph and the following plane models using the square, the double pyramid, and the 5-cell.

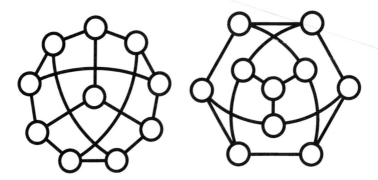

Of course, syntheme–duad geometries can be defined on any number of points. Later we will see that projective planes of order n that contain an oval or hyperoval can be regarded as generalized syntheme–duad geometries over the oval or hyperoval, respectively. Without going into the details right now, we state the following construction principle for "good pictures" of (generalized) syntheme–duad geometries:

Construction principle: syntheme–duad geometries

Given an abstract syntheme–duad geometry on n points, try to translate automorphisms of symmetric arrangements of n points in the plane or in space into "good" models of the geometry that exhibit as many of these automorphisms as possible.

4.4   The Generalized Quadrangle of Order $(2, 4)$

There is a unique generalized quadrangle of order $(2, 4)$ that contains the generalized quadrangle of order $(2, 2)$ as a subgeometry. We first extend the abstract description of the generalized quadrangle of order $(2, 2)$ in terms of duads and synthemes to an abstract description of the generalized quadrangle of order $(2, 4)$: In addition to the duads and synthemes, let $1, 2, \ldots, 6$ and $1', 2', \ldots, 6'$ denote twelve additional points, and let $\{i, ij, j'\}$, $1 \leq i, j \leq 6$, $i \neq j$, denote thirty additional lines. Then the 27 points and 45 lines just constructed form a geometry isomorphic to the unique generalized quadrangle of order $(2, 4)$.

We already used the following construction principle for pretty pictures of geometries once in the section on the Petersen graph when we extended the most famous representation of this graph to a picture of the generalized quadrangle of order $(2, 2)$.

> **Construction principle: subgeometry → full geometry**
>
> Try to extend "good" models of subgeometries of a given geometry to a "good" model of the full geometry.

The good model that we are going to use in the following is the doily. We first label the doily with duads and thereby establish an isomorphism of the abstract model and the pictorial model of the generalized quadrangle of order $(2, 2)$. Here is how we add the additional 12 points to our diagram. Note that the points 6 and 6′ are both located in the middle of the diagram, or rather, they are located at equal distance above and below the plane the doily is drawn in. You have to merge the following two pictures in your mind to get a good three-dimensional picture.

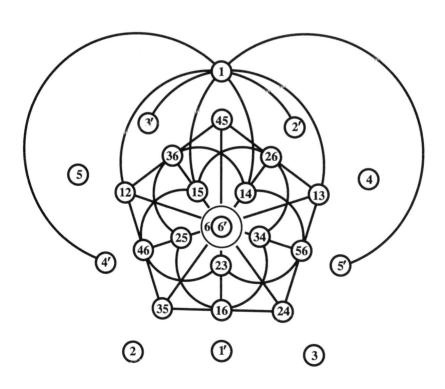

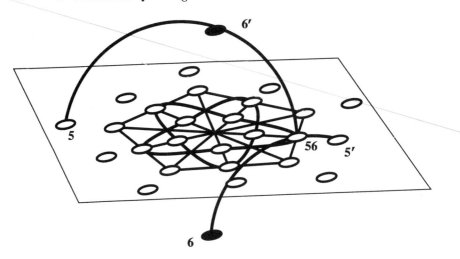

This generalized quadrangle also contains spreads consisting of 9 lines each. Try to find one.

Here is the part of the geometry that lives in the same plane as the doily.

4.5   The Generalized Quadrangle of Order $(4, 2)$

There is a unique generalized quadrangle of order $(4, 2)$, which again contains the generalized quadrangle of order $(2, 2)$ as a subgeometry. It is the dual of the generalized quadrangle of order $(2, 4)$ that we considered in the last section.

We again extend the doily. We first add 30 points to the diagram, 2 on each line.

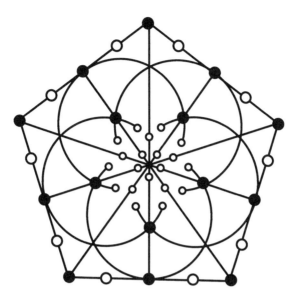

And here are the additional lines that have to be added. Note that two of these additional lines are represented by circles in the middle of the diagram.

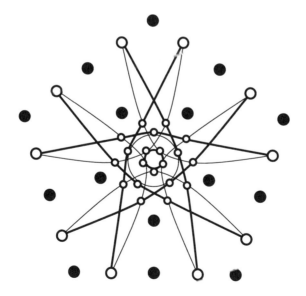

On closer inspection it turns out that the 30 additional points plus the 12 additional lines form an abstract plane configuration with parameters $(30_2, 12_5)$. This configuration is also known as *Schläfli's double six*. It is nothing but a 6×6 grid with the points on the diagonal missing. See [49] for a good introduction to this important configuration.

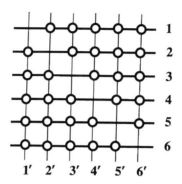

There is a very nice spatial representation of this double six on the cube.

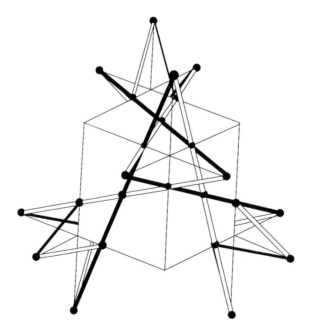

If we project this three-dimensional model in the direction of the diagonal of the cube through the vertex closest to us onto a plane perpendicular to this diagonal, we arrive at the following picture.

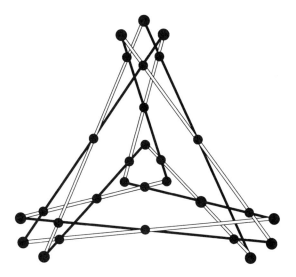

The three-dimensional model can be extended to a quite fantastic model of the generalized quadrangle of order $(4, 2)$ that uses only straight-line segments as lines and that is also an extension of our generalized quadrangle of order $(2, 2)$ on the tetrahedron. The following stereograms show how to construct this model starting with the "extended cube." The first stereogram shows this extended cube, the second one the extended cube plus all the points and lines of the double six. The third one shows how to construct the additional 15 points and lines starting with the points and lines of the double six; the fourth stereogram shows the generalized quadrangle of order $(2, 2)$ formed by these 15 points and lines; the fifth stereogram shows the full model. Note that Appendix B contains a larger version of the fifth stereogram that can be viewed with the mirror technique.

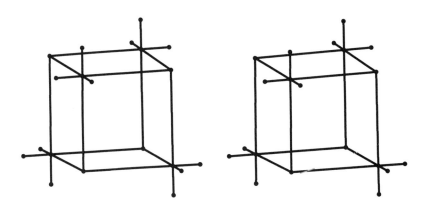

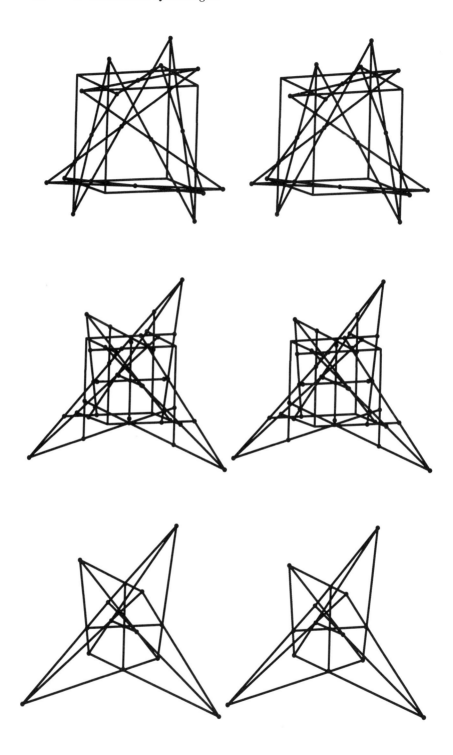

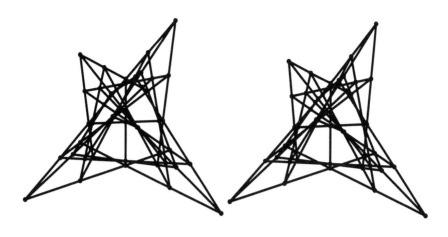

4.6   Symmetric Designs and Generalized Quadrangles

Given any generalized quadrangle of order (q, q), it is possible to construct a symmetric $2 - (q^3 + q^2 + q + 1, q^2 + q + 1, q + 1)$ design as follows: The points of the design are the points of the quadrangle. Associated with every point is a block consisting of all those points that are collinear with the point. For $q = 2$, we get a $2 - (15, 7, 3)$ design that turns out to be the geometry of subplanes of the smallest three-dimensional projective space, which we are going to consider in Chapter 5. This $2 - (15, 7, 3)$ design is also an example of a Hadamard 2-design (see Section 2.2).

Given any generalized quadrangle of order $(q - 1, q + 1)$, it is possible to construct a symmetric $2 - (q^2(q + 2), q(q + 1), q)$ design as follows. Take as points of the design the lines of the quadrangle. To any line in the quadrangle associate the set of lines that intersect the line and are different from the line. These sets form the blocks of the design. In the case of the generalized quadrangle of order $(2, 4)$, we arrive at a symmetric $2 - (45, 12, 3)$ design.

More details about these constructions can be found in [74, pp. 59–65].

4.7   Incidence Graph of a Generalized Quadrangle

Here is a picture of the incidence graph of the generalized quadrangle of order $(2, 2)$. By now it should be clear what the labelling is all about. For more details about this graph see Chapter 13, on generalized polygons. Note that 5 of the 6 polarities of the generalized quadrangle correspond to

5 reflections of the diagram. Can you see the corresponding 5 ovoids and spreads, and can you find the missing polarity in the diagram?

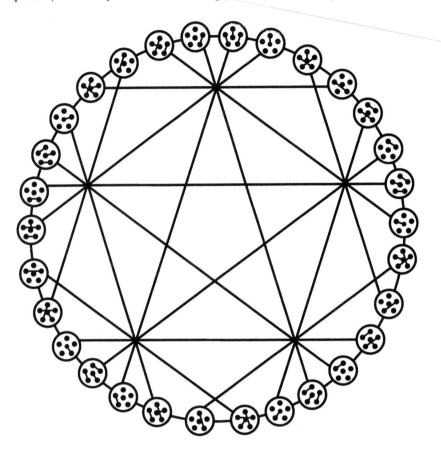

5

The Smallest Three-Dimensional Projective Space

We call a set of three lines in a geometry a *triangle* if each pair of lines in the set intersect in a point and the three points in which the lines intersect are distinct. We call the three lines the *sides* and the three points of intersection the *vertices* of the triangle. A geometry is called a *projective space* if it satisfies the following three axioms:

Axioms for projective spaces

(PS1) Two distinct points are contained in a unique line.

(PS2) A line that intersects two sides of a triangle but does not contain any of the vertices of the triangle also intersects the third side of the triangle.

(PS3) Every line contains at least three points.

Here is a pictorial version of the second axiom. The three solid black lines and their points of intersection form the triangle. Note that the configuration on the right is a complete quadrangle, that is, an APC with parameters $(6_2, 4_3)$ (see also p. 35).

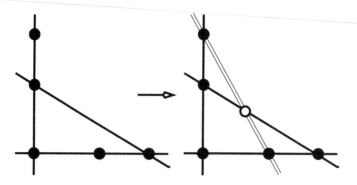

The lines are the *1-dimensional subspaces* of a projective space. For $m > 1$, the m-dimensional subspaces are constructed from the $(m - 1)$-dimensional subspaces as follows: Take an $(m - 1)$-dimensional subspace and a point not contained in it. Then the m-dimensional subspace *generated* by the point and the subspace is the union of all lines connecting the point with points in the subspace.

If an m-dimensional subspace coincides with the whole space, then all m-dimensional subspaces do so, and m is called the *dimension* of the projective space under consideration. It turns out that the 2-dimensional projective spaces are just the projective planes. An m-dimensional subspace of a projective space, together with all the lines contained in it, is an m-dimensional projective space. As geometries, all m-dimensional subspaces of a finite-dimensional projective space are isomorphic. The 2-dimensional subspaces of a projective space have a special name. They are called *subplanes*. In finite projective spaces, all subspaces of a given dimension contain the same number of points. If $n + 1$ is the number of points in a line, the number n is called the *order* of the projective space. Given an integer $m > 2$ and any prime power n, there is a unique m-dimensional projective space of order n. It is usually denoted by $\mathrm{PG}(m, n)$. The geometry we arrive at by removing an $(m - 1)$-dimensional subspace plus all its points from $\mathrm{PG}(m, n)$ is an *m-dimensional affine space of order n*. Again, there is a unique m-dimensional affine space of order n. It is denoted by $\mathrm{AG}(m, n)$.

In this section we present a number of models for the smallest three-dimensional projective space of order 2, that is, $\mathrm{PG}(3, 2)$.

First, let us consider the simple geometry $\mathrm{AG}(3, 2)$. A simple model of this geometry is the complete graph on the 8 vertices of a cube. Note that there are 14 copies of the affine plane $\mathrm{AG}(3, 2)$, which is just the complete graph on 4 points, embedded in this space.

For more information about finite affine and projective spaces see [51]. A large part of this chapter can be regarded as a pictorial version of [11], [59], [60], and [66], which, by the way, can serve as highly accessible introductions to $\mathrm{PG}(3, 2)$.

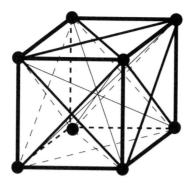

5.1   A Plane Model

The geometry $PG(3,2)$ has 15 points, 15 subplanes, and 35 lines; every line contains 3 points, and every point is contained in 7 lines. Furthermore, three noncollinear points are contained in a unique subplane, and two subplanes meet in precisely one line.

On page 25 we considered a generator-only model of a Steiner triple system of order 15. This Steiner triple system is the geometry of points and lines of the projective space we are after in this section. Note that there is a Singer diagram of the Fano plane sitting right in the middle of this model. Its points are the open points, and its lines are generated by the triangle all of whose vertices are open points. Try to find the remaining 14 subplanes in this model.

The generalized quadrangle of order $(2,2)$ is naturally embedded in the projective space $PG(3,2)$ as the geometry whose points are the points of $PG(3,2)$ and whose lines are the totally isotropic lines in $PG(3,2)$ provided with a nondegenerate symplectic form (see [74]). In Sections 4.4 and 4.5 we extended the doily to nice models of the generalized quadrangles of orders $(2,4)$ and $(4,2)$. Now we want to extend it to a model of $PG(3,2)$.

We already mentioned in Section 4.6 that the symmetric $2 - (15,7,3)$ design associated with the generalized quadrangle of order $(2,2)$ is the geometry of subplanes of the space we are after in this section. This gives a way in which we can construct the space around any model of the generalized quadrangle. First, we construct the $2 - (15,7,3)$ design. This yields all the subplanes of the space. Then we construct all possible intersections of subplanes. This gives all the lines. Alternatively, let x and y be two different points of the quadrangle and let $x^\perp$ denote the set consisting of the 7 points collinear with x, and $\{x,y\}^\perp$ the set consisting of the 3 points collinear with both x and y. Then the lines of our space are all possible sets $\{x,y\}^\perp$, and the planes are all possible sets $x^\perp$. In this way, we find how a generator-only picture of the generalized quadrangle can be extended to a generator-only picture of the three-dimensional space.

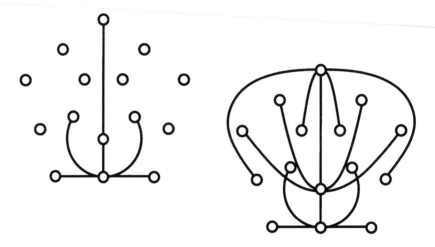

5.1.1 Line Pencils

Here are the different line pencils (up to rotation). There are 15 line pencils, corresponding to the 15 points of the space. Note that given a point of the space, the line pencil through this point partitions the set of points of the space other than the point. This means that as expected, two points of the space are contained in exactly one line. Furthermore, it is clear from the way we constructed the space that two planes intersect in exactly one line of the space. Note that $\{x, y\}^{\perp} = x^{\perp} \cap y^{\perp}$.

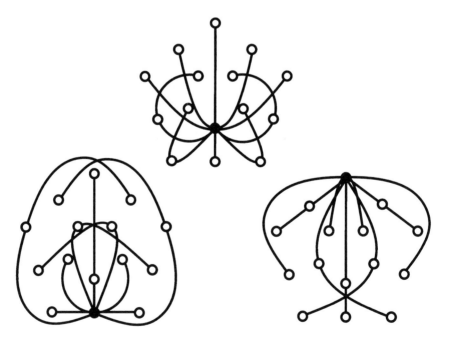

5.1.2   Subplanes

The space has 15 subplanes. Here they are (up to rotation of the diagrams).

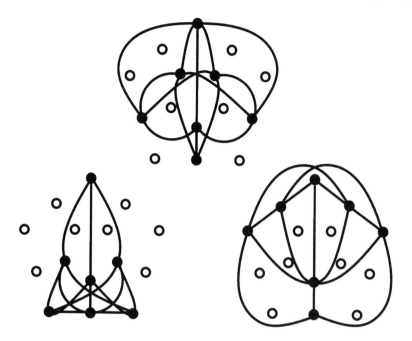

Check that all these subplanes are indeed Fano planes and that any two such subplanes have exactly one line in common. Note also that every subplane intersects the quadrangle in exactly 3 lines.

5.1.3   Spreads and Packings

Spreads

A *spread* in our projective space is a set of five lines partitioning the point set. There are 56 such spreads contained in the space (see [51, p. 55]). Here they are (up to rotation). Note that every spread except the one in the middle of the first row stands for 5 spreads.

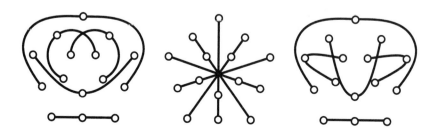

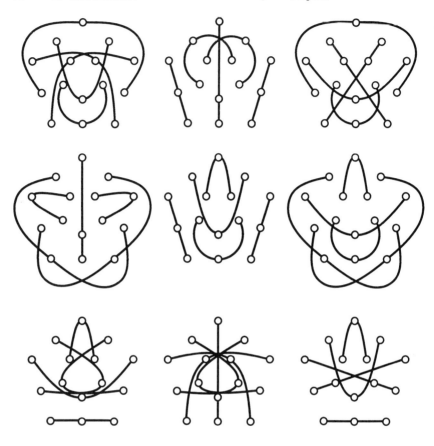

Two lines in the projective space are called *skew* if they do not intersect. Here are a couple of interesting facts that you might want to check using our list of spreads:

- Given any two skew lines of our space, there are exactly two spreads containing both of them.

- Every set of three or four mutually skew lines is contained in exactly one spread.

- For three mutually skew lines, there are only two lines skew to all three and only three lines intersecting all three.

Packings

A *packing* of the projective space PG(3, 2) is a set of seven pairwise disjoint spreads. This implies that every line in PG(3, 2) is contained in exactly one spread of a packing. There are 240 different packings of PG(3, 2) (see [51, Thm. 17.5.6]). A packing of the projective space is also a solution of

Kirkman's schoolgirl problem (see Section 2.3 for more details about this interpretation).

Here is an example of a packing.

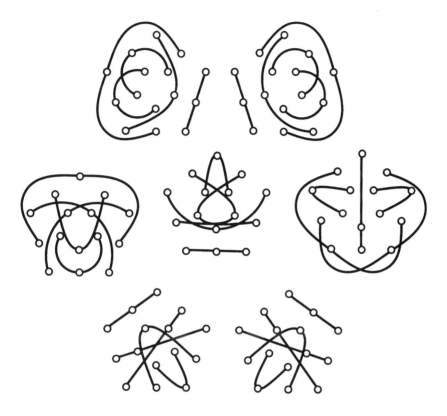

Check that every line of our space is indeed contained in exactly one of the spreads in this particular packing.

Fano Planes Everywhere

Given a packing like the one above, it is possible to turn it, in a natural way, into a Fano plane with the 7 spreads corresponding to the 7 Points (with a capital "P") of the plane. Here is the rule that we use to construct the Lines (with a capital "L") of the Fano plane: Take any two Points/spreads X and Y. Then there is exactly one line l in the projective space forming a spread W with two lines from X and Y each. This line is contained in a unique spread Z in the packing under consideration, and X, Y, and Z form a Line in the Fano plane. This means, of course, that if we start with X and Z, for example, we end up constructing Y.

We label the spreads in the above packing from left to right and top to bottom with the numbers $1, 2, 3, 4, 5, 6$, and 7. We want to construct the third point on the line connecting the Points $X = 1$ and $Y = 2$. Here is the line l and the spread W.

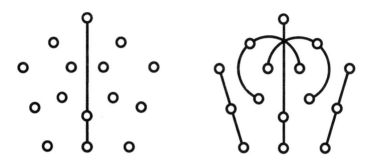

Since the line l is contained in Point 5, this is the Point we have been looking for. If we continue to construct the remaining lines of the Fano plane in this manner, we arrive at the following diagram, which shows all the Lines.

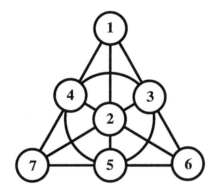

N-Spreads and a Desargues Configuration

An *n-spread* of lines in our space is a set of $5n$ lines such that every point in the space is contained in exactly n lines in the set. Therefore, the spreads are just the 1-spreads. Also, the union of n disjoint spreads is an n-spread. We call an n-spread *irreducible* if it does not contain any m-spreads with $m < n$.

Here is a construction for irreducible 2-spreads. Take any generalized quadrangle of order $(2, 2)$ embedded in our projective space and remove from its line set the lines of a spread that is contained in the generalized quadrangle. Then the remaining 10 lines form an irreducible 2-spread. The picture on the left is an example of such a 2-spread. All 2-spreads are isomorphic to this one.

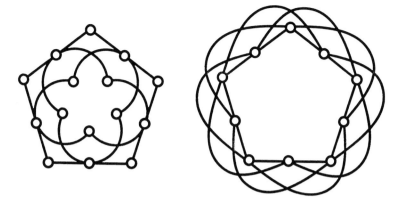

We construct a graph as follows: The vertices are the lines in the 2-spread, and two vertices are connected by an edge if and only if the corresponding lines in the 2-spread intersect. This graph is the Petersen graph, and for this reason, our 2-spreads are also called *Petersen systems*. There are 168 Petersen systems contained in the projective space (see [60, Theorem 5]).

The picture on the right above is a Desargues configuration. Just like the picture of the Petersen system this picture is generated by two lines. The classical projective planes are characterized among the projective planes by the fact that all Desargues configurations close. In projective spaces of finite dimension greater than 2, all Desargues configurations close, and in fact, all these higher-dimensional projective spaces are "classical."

5.1.4   Reguli

A set of three pairwise skew lines in our projective space is called a *regulus*. There is a natural pairing up of reguli where two reguli form a pair if and only if every single one of the lines in one intersects all the lines in the other. Here are two such pairs. Try to find some more. You should not have any problems with this exercise, since there are 560 reguli to choose from (see [51, p. 55]).

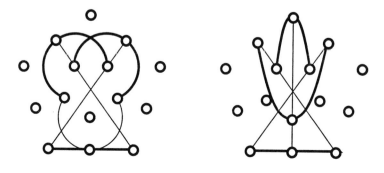

The nine points of intersection in any regulus like this are the points of a *hyperbolic quadric*. All except 6 planes in the projective space intersect this set of 9 points in a pair of lines. Define a geometry whose point set consists of the 9 points and whose line set consists of the 6 sets of intersection of the 6 distinguished planes. Then this geometry is the Minkowski plane of order 2 (see Chapter 12 for more information about Minkowski planes).

5.1.5   Ovoids

We continue with a selection of *ovoids*, that is, sets of 5 points each such that a line in our space intersects the set in at most 2 points. The following diagram shows a partition of our space into 3 ovoids that correspond to the 3 concentric regular pentagons that are hiding in the diagram. There are 168 different ovoids contained in our projective space. Every ovoid is an *elliptic quadric*.

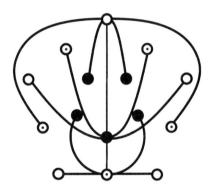

Any subplane of our space intersects a given ovoid in either 1 or 3 points. In the first case it is called a tangent plane, in the second case a secant plane. We concentrate on the ovoid consisting of the open points. Going back to our list of subplanes, we see that the five planes represented by the first of the three diagrams in our list are the tangent planes. All other planes are secant planes. The *inversive plane* associated with the ovoid is defined as follows: As points choose the points of the ovoid and as circles choose the intersections of the secant planes with the ovoid. These intersections are just the 10 different subsets of 3 elements each of the point set, that is, all possible triangles having three of the open points as vertices.

The resulting geometry is the trivial $3 - (5, 3, 1)$ design, also called the *inversive plane* of order 2. The derived geometry at a point of this inversive plane is the affine plane of order 2. See also Chapter 12 for more information about inversive planes.

5.1.6 A Labelling with Fano Planes

A set of 7 labels can be assembled into Fano planes in 30 different ways. Two such labelled Fano planes have 0, 1, or 3 lines in common. There is a unique way to divide up this collection of 30 Fano planes into two parts X_1 and X_2 containing 15 planes each such that two distinct planes both in X_j have precisely one line in common ($j = 1, 2$). Now PG$(3, 2)$ can be constructed as follows: Take as points the Fano planes in X_1, as lines the 35 different subsets of 3 each of the 7 labels, and as planes the Fano planes in X_2 (with the obvious incidence relations). In order to avoid confusion, let us refer to the points and lines of the projective space as "Points" and "Lines" and to points and lines of the Fano planes just as usual as "points" and "lines."

The following diagram shows a labelling of the Point set of our projective space via labelled Fano planes. The labelling we use is the one derived in [66] (see also [44]).

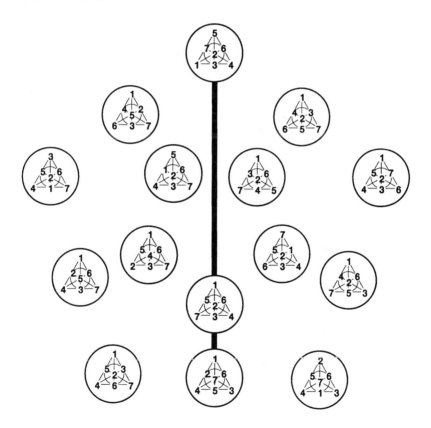

For example, the highlighted vertical line in the diagram corresponds to the Line 157.

We use the remaining set of 15 labelled Fano planes to label the 15 Planes of our space. We arrive at a diagram similar to the one above. A polarity π of our space can be defined by exchanging Points and Planes that occupy respective positions in the two diagrams. In fact, given a plane P in our space, the Points contained in it are just the Points collinear with $\pi(P)$ in the doily. Also, these 7 Points are exactly the Points that have exactly 3 lines, that is, 3 triples of labels, in common with the labelled Fano plane that denotes the Plane P.

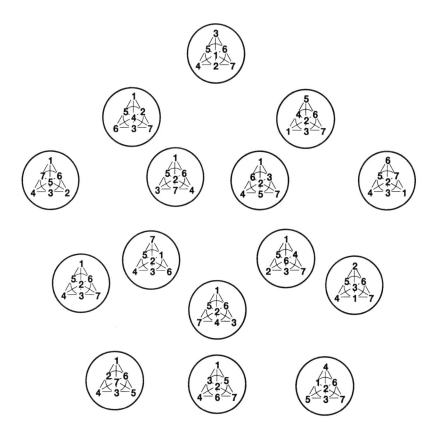

Let i be one of the labels. Then the set of all Points of our space together with the set of all Lines that contain this label is a generalized quadrangle of order $(2, 2)$. It turns out that the 7 labels correspond to 7 of the generalized quadrangles of order $(2, 2)$ embedded in the projective space.

By letting i equal 7, we arrive at the generalized quadrangle around which we built the projective space. Here is the corresponding labelling of the lines in this quadrangle (we always omit the 7). Note that this is a labelling of the lines by duads (Section 4.1.3) and that the duads corresponding to the three lines through a point of the quadrangle always form a syntheme.

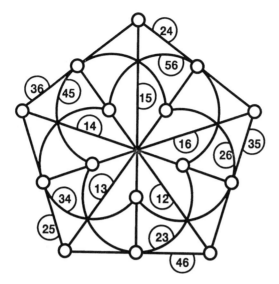

Note that the 5 lines containing two fixed labels i and j always form a spread. The 6 spreads of the generalized quadrangle that corresponds to the label 1 are the images under rotation of the following two spreads.

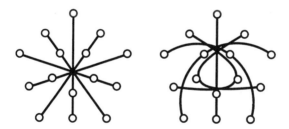

Here are the 6 spreads of the generalized quadrangle that corresponds to the label 5.

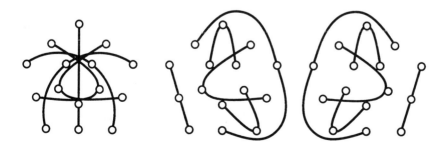

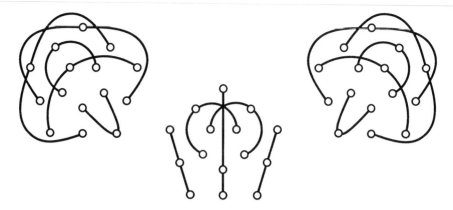

We arrive at the generalized quadrangles that correspond to the remaining 4 labels by simultaneously rotating the above six spreads.

Exactly 21 of the 56 spreads in our projective space arise in this manner. The remaining 35 arise as follows: Take any triple of labels plus the four triples of labels disjoint from it. Then the five lines corresponding to these 5 triples form a spread.

E. van Dam (see [113]) describes the following natural one-to-one correspondence between the set of spreads and the set of unordered triples of labels taken from the set consisting of the numbers from 1 to 8: Given a spread of the first kind, all triples corresponding to its lines have two labels x and y in common. We let this spread correspond to the triple $\{x, y, 8\}$. We let a spread of the second kind correspond to the distinguished triple contained in it. Now, a resolution of our projective space corresponds to a set of 7 triples all of whose mutual intersections have size one. In this way we rediscover that such a packing can be made into a Fano plane in a natural way: Let the points be the seven labels of intersection and the lines the seven distinguished triples.

Note also that two of the seven generalized quadrangles corresponding to labels intersect in exactly one spread and that any line in one of the generalized quadrangles is contained in exactly two of its spreads.

5.1.7   Fake Generalized Quadrangles

We have already come across a fake generalized quadrangle in Section 4.1.2, that is, an abstract plane configuration with parameters (15_3). Here are generator-only pictures of two more fake generalized quadrangles in $\mathrm{PG}(3, 2)$. Together with the spread on the right they form a remarkable partition of the line set into two fake generalized quadrangles and a spread, that is, a kind of packing different from the one we considered before. The next two pictures show a triangle in the two geometries, demonstrating that we are indeed not dealing with generalized quadrangles.

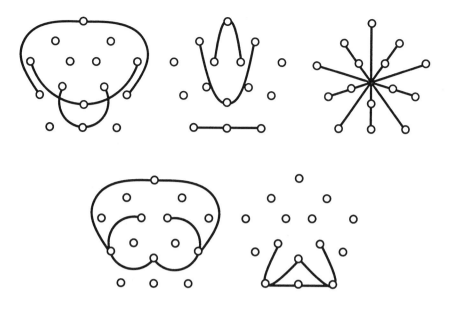

By the way, there exist 4 different types of fake generalized quadrangles in our projective space (see [59]).

5.1.8   The Hoffmann–Singleton Graph

Using the particular labelling of the Points of the projective space it is possible to construct the so-called *Hoffmann–Singleton graph* (see [18, p. 391]). This graph is a very important relative of the Petersen graph. It has 50 vertices and is a regular graph of valency 7. Furthermore it is:

- the largest known Moore graph (see [24]);

- the unique $(7, 5)$-cage (see Section 13.6).

Here is a construction of this graph (another one is given in Section 13.6): The vertices of the graph are the 15 Points and the 35 Lines of $PG(3, 2)$. Two Points of $PG(3, 2)$ are never adjacent in the graph; a Line and a Point of $PG(3, 2)$ are adjacent if the Point is contained in the Line; and two Lines, that is, two sets of 3 labels, are adjacent if they are disjoint. The following diagram shows part of the graph. It illustrates that the graph is of valency 7 and that there are circles of length 5 in this graph. Try to convince yourself that 5 is the girth of the graph, that is, all circles in the graph are of length at least 5. Also try to verify that the diameter of this graph, that is, the maximum distance between two vertices, is 2. Finally, try to find a subgraph that is isomorphic to the Petersen graph.

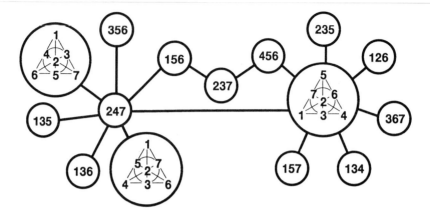

As we said before, the Hoffmann–Singleton graph is closely related to the Petersen graph. Here is a construction of the Petersen graph that is similar to the construction that we just considered: Let the vertices of the Petersen graph be the points and lines of the affine plane of order 2. Two vertices that are points are never connected by an edge. A vertex that is a point gets connected to a vertex that is a line if and only if the point is contained in the line. Two vertices that are lines are connected by an edge if and only if they are parallel.

Starting with the picture of the affine plane on the square, try to come up with a meaningful labelling of our spatial model of the Petersen graph.

5.2 Spatial Models

5.2.1 A Model on the Tetrahedron

There is an amazing model of PG(3, 2) that was first shown to me by A. Offer, a Ph.D. candidate at the University of Adelaide. Later on I discovered the following two references for this model: [43], [66]. Some of my colleagues believe that D. Mesner was the first to discover it. The points of the model are the 4 vertices of the tetrahedron, its center, the 4 centers of its faces and the 6 centers of its edges. Its lines are the 6 edges of the tetrahedron, the 12 medians of its faces, the 4 circles inscribed in the faces, the 3 segments connecting opposite edges of the tetrahedron, the 4 medians of the tetrahedron, and 6 circles that are situated inside the tetrahedron. These 6 circles correspond to the 6 edges of the tetrahedron. A circle like this touches one of the edges in its center and also contains the centers of the two faces "opposite" this edge, that is, the faces that have precisely one point each in common with the edge. Here is a stereogram picture pair of our projective space created by A. Schroth. You can view it with both the cross-eyed technique and the parallel technique.

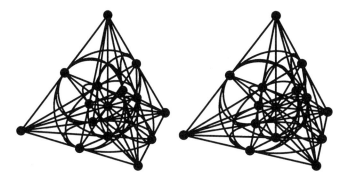

See also in Chapter 14 the photo of this model built from pipe cleaners. It was such a pipe cleaner model which first sparked my interest in writing this book.

Line Pencils

Here are the four different line pencils in the model corresponding to the four different kinds of points.

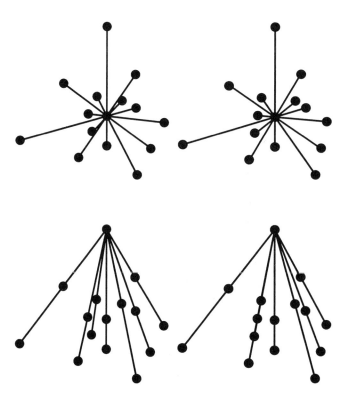

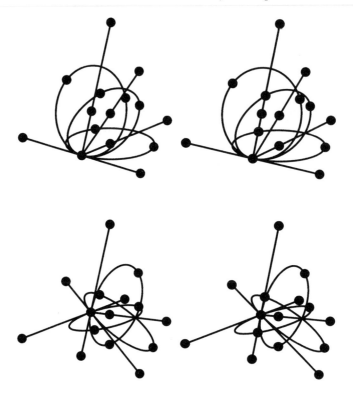

Subplanes

There are essentially 4 different models of subplanes embedded in this
model. The first diagram stands for one subplane, the second for 4, the
third for 6, and the fourth for 4 subplanes.

Note that this subplane corresponds to our first model of the Fano plane
on the tetrahedron discussed in Section 1.10.

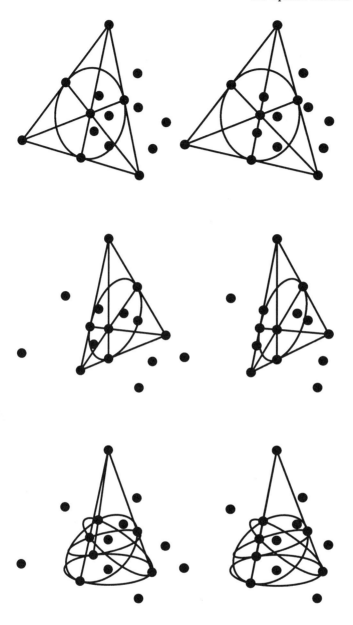

Remember that the Fano plane can also be constructed as follows: Fix a point p in our projective space. Let the points and the lines of the Fano plane be the lines and the subplanes containing p, respectively. By choosing the point p to be the center of the tetrahedron, we arrive at the second spatial model of the Fano plane that we introduced in Section 1.10.

5.2.2   Other Substructures

The following stereograms show a model of the complete quadrangle, that is, the PC with parameters $(6_2, 4_3)$; the tetrahedron, which is the dual of the complete quadrangle; a Desargues configuration, which turns out to be the union of a complete quadrangle and its dual; and, finally, the generalized quadrangle of order $(2, 2)$.

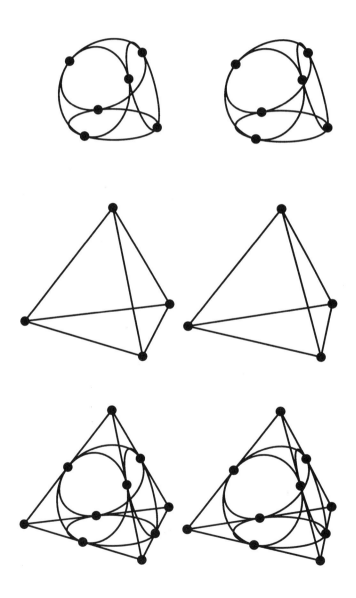

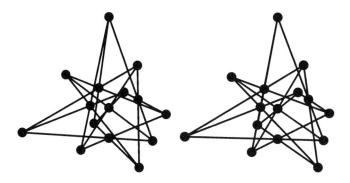

If we start with this generalized quadrangle and apply the extension process described at the beginning of this chapter to it, we arrive at the model of the projective space.

Here is a highly symmetric fake generalized quadrangle embedded in our model. Take as points the points of the model of $PG(3,2)$ and as lines the 6 edges of the tetrahedron, the 6 circles inscribed in the model, and the 3 line segments connecting opposite edges of the tetrahedron. In fact, on closer inspection this fake generalized quadrangle turns out to be isomorphic to the one on the cube that we described in Section 4.1.2. In order to see this, note that in both cases the line set is the union of two tetrahedrons together with a tripod of lines through the respective center of the models.

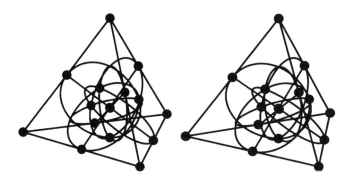

Finally, here is another picture of the Petersen graph right in the middle of the generalized quadrangle. We construct it by removing the points of an ovoid from the above generalized quadrangle. For more information about this see page 51.

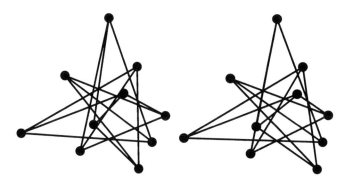

5.2.3 Spreads

The model contains 5 essentially different spreads. Via rotations and re-flections of the tetrahedron of reference, the following spreads generate 6, 6, 8, 12, and 24 spreads, respectively.

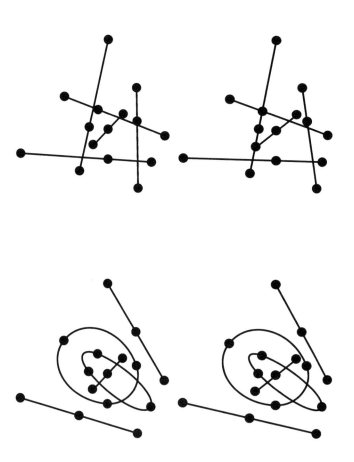

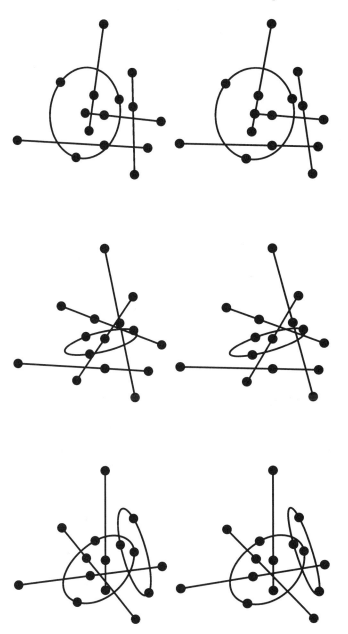

Here is a picture of a 2-spread embedded in our space. Note that the generalized quadrangle above is the union of this 2-spread and the first of the above spreads.

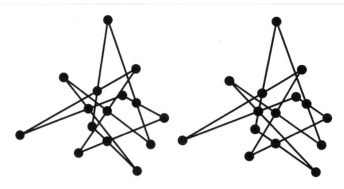

5.2.4   A Model on the Icosahedron

It is also possible to extend our model of the generalized quadrangle of order $(2, 2)$ on the icosahedron to a model of the projective space.

Here again are the generating lines for the generalized quadrangles representing 5 and 10 lines, respectively. These are followed by two more lines that generate the remaining lines in the projective space. Each of these two lines represents 10 lines in the space.

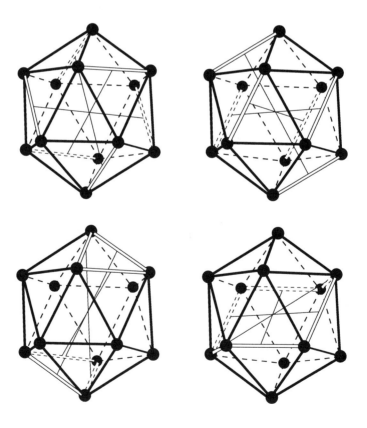

In [96] Shaw gives a very nice representation of PG$(5, 2)$ on the icosahedron that contains our model "right in the middle."

5.3 Symmetric Designs Associated with Our Space

We already mentioned in Section 2.2 that the geometry of subplanes of PG$(3, 2)$ is a Hadamard design with parameters $2 - (15, 7, 3)$. There is another interesting construction of a Hadamard design with parameters $2 - (35, 17, 8)$ from this space (see [58] and [18, p. 227 (iii)]): Take as points and blocks of the design the lines of the space, and let a point be incident with a block if they coincide or if they are disjoint as lines of the space.

6

The Projective Plane of Order 3

6.1 More Models of the Affine Plane of Order 3

In addition to the traditional picture of the Fano plane, the following picture of the affine plane of order 3 has been a favourite with authors of textbooks on incidence geometry.

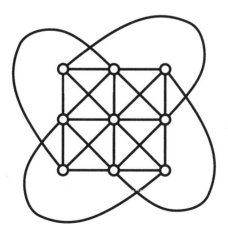

We have already come across two other pictures of the affine plane of order 3: a nice generator-only model on the regular octagon on page 24 and one built around the traditional picture of the Pappus configuration on page 37.

The next picture shows the different parallel classes, again, only up to rotational symmetry. And below are the different line pencils.

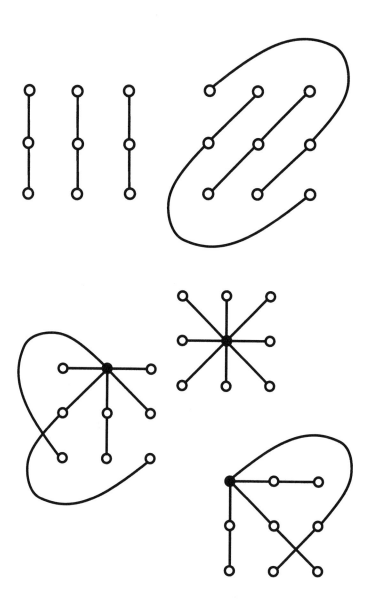

6.1.1   Two Triangular Models

Here is another nice model of the same geometry (see [61]) together with a picture of the parallel classes in this model.

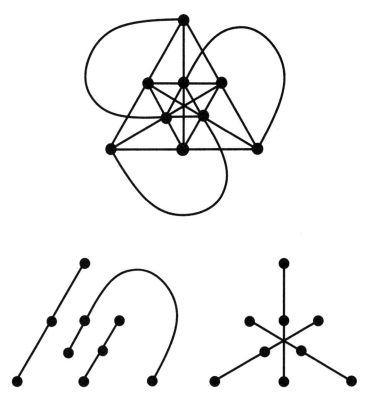

Another Pappus Model

In Section 3.3 we constructed our affine plane around a picture of the Pappus configuration. Here is another highly symmetric picture of this configuration.

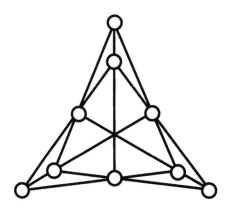

And here is how 3 lines have to be added to this picture to make it into a picture of the affine plane of order 3.

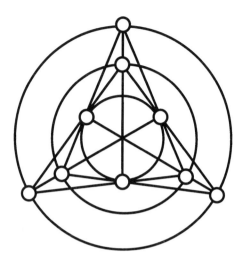

6.1.2 Maximizing the Number of Straight Lines

Here again are the five pictures of our affine plane that we have considered so far.

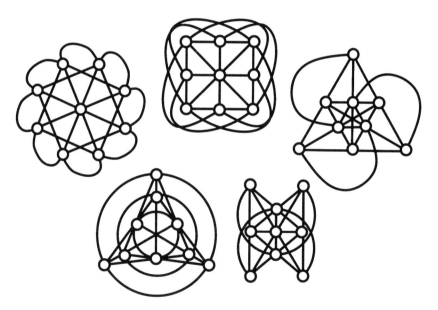

In drawing these pictures we used 4, 8, 9, 9, and 10 straight-line segments as lines of the plane, respectively. It does not seem to be possible to do better than 10 if one tries to draw a picture of this plane in this straightforward way.

There is a way to draw this geometry using only straight lines. In fact, this is possible for every single one of the affine planes of prime order: Start with the following infinite grid. Note that this grid is the 3×3 grid in the middle of the picture repeated in a periodic manner to fill in the whole plane.

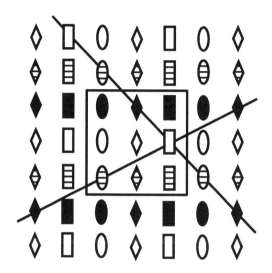

The points of the affine plane are the nine subgrids corresponding to the different symbols in the 3×3 grid in the middle of the picture. In order to construct the connecting line of two points/subgrids, just take one grid point from each of the two subgrids and connect them by a straight line. This can usually be done in several different ways as in the example above, but no matter which of the ways you choose, you always end up with exactly the same three different labels on this line: the two we started out with and a third one that corresponds to the missing third point on the connecting line of the two points we started out with. In this way, all lines of the affine plane of order 3 can be constructed with straight lines. The same construction also works for all other classical planes of prime order p; just replace the 3×3 grid in the middle of the picture by a $p \times p$ grid consisting of $p \times p$ different symbols. Of course, what we do here is just a straightforward reinterpretation of the affine plane of order p represented as the geometry whose point set is the grid $\mathbf{Z}_p \times \mathbf{Z}_p$ and whose lines are the verticals in this grid plus the graphs of all linear functions over $\mathbf{Z}_p$. The traditional picture of the affine plane is another drawing of the same representation.

6.2   Projective Extension

The following variation of one of our models of AG(2, 3) can be extended to a model of PG(2, 3) in a straightforward way.

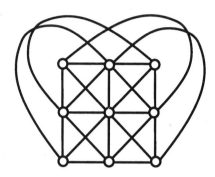

Here is the picture of this extension by the (thick) line at infinity. Note that members of the same parallel class get extended by the same (solid) *point at infinity*. This picture appears in a number of textbooks.

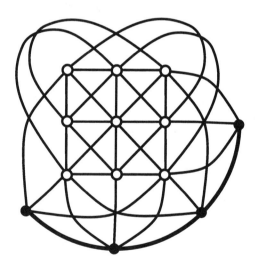

The picture itself is not symmetrical. We can create a symmetrical picture by extending the traditional model of AG(2, 3) by a line at infinity that is covered twice. In the following picture this line at infinity is again drawn as a thick circle, and every pair of antipodal points on this circle corresponds to one point at infinity. Let the center of the generator-only picture on the left be the origin of the xy-plane. Starting with this picture, we construct the full picture on the right by rotating around the origin, 90 degrees at a time, and reflecting in the coordinate axes.

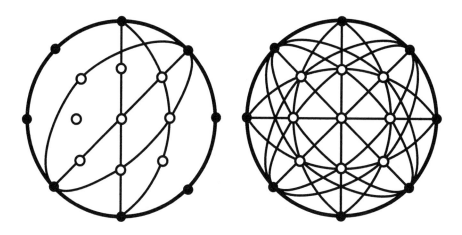

Another model that extends the model on the 3×3 grid is that by Lehti [64]: The line at infinity is the thick line on top, and the remaining lines are either straight-line segments or four "broken" lines made up of two straight-line segments that meet in a point at infinity. The four short brackets indicate in what way these broken lines are spliced together.

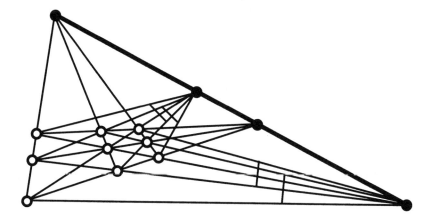

Finally, we extend our first triangular picture of $AG(2,3)$ to a picture of $PG(2,3)$.

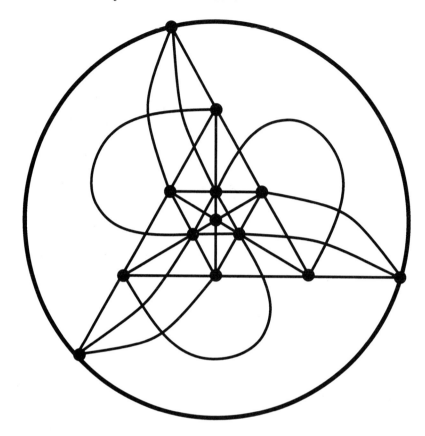

Note that the point in the middle of the diagram is supposed to be contained in the "circle."

6.2.1 Blocking Sets

Remember that a blocking set in a geometry is a set of points such that every line of the geometry contains at least one point in the set (see Section 1.7). What are examples of minimal blocking sets in our projective plane? In [50, Lemma 13.1.5 (i)] we find the following construction of minimal blocking sets in PG$(2, q)$, with $q > 2$: a line l minus a point P plus a set of q points, one on each of the q lines through P other that l but not all collinear. This minimal blocking set contains $2q$ points. In the following example the arrow points at l. The set of open circles is the blocking set. Convince yourself of this fact. Also check that this blocking set is minimal by verifying that through every point of the set there is a line containing only this point.

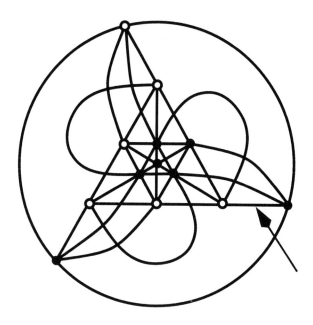

6.2.2   Desargues Configuration

The following Desargues configuration sits right in the middle of our diagram, showing that this configuration admits an automorphism of order 3.

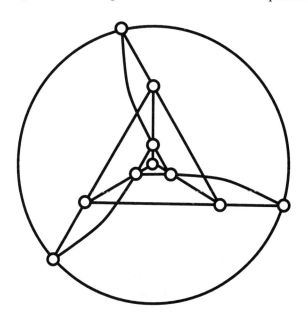

6.3   Singer Diagram and Incidence Graph

We draw a picture of a partial Singer diagram (it should be clear how to complete it) and the incidence graph of PG(2, 3) as described in Sections 1.8 and 1.9. A difference set of order 13 corresponding to PG(2, 3) is {0, 1, 3, 9} (see [3, p. 150]).

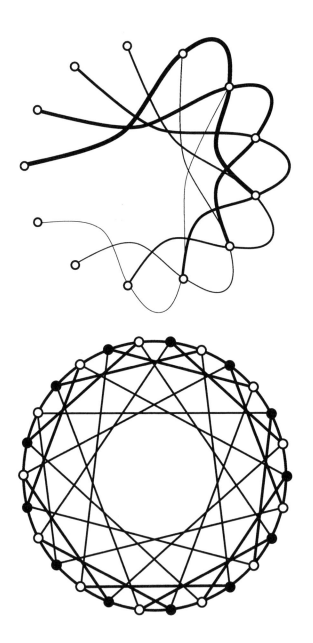

6.4   A Spatial Model on the Cube

The following spatial model lives on the cube: The points of the plane are the 13 rotation axes of the cube, or equivalently, the 13 pairs of points on the cube that correspond to the pairs of points of intersections of the axes with the surface of the cube. The following pictures show the three essentially different lines in this model consisting of four axes each. The first line stands for 6 lines in the model, the second one for 3, and the last one for 4.

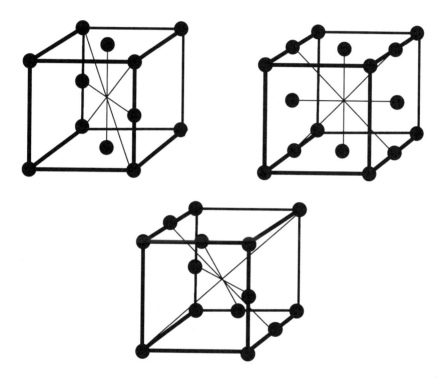

Try to visualize viewing this model from "above." Can you see our model of the projective plane on page 99 with the line at infinity covered twice?

This model has counterparts for the projective planes of orders 2 and 5 on the tetrahedron and the dodecahedron, respectively (see Sections 1.10 and 8.2).

The octahedron, too, has 13 rotation axes, and a model of this projective plane can also be defined in terms of these rotation axes. Just remember that the octahedron can be inscribed in the cube such that the vertices of the octahedron are the centers of the faces of the cube. Inscribed in this way, the rotation axes of the octahedron and the cube coincide.

A closer look at the different kinds of points and lines in the model on the cube yields the following neat interpretation:

- The 4 axes through the vertices of the cube form an oval.

- The 3 axes through the centers of the faces and the 6 axes through the centers of the edges are the 3 interior and the 6 exterior points of the oval.

- The lines of the first, second, and third kinds correspond to the secant, exterior lines, and tangent lines of the oval.

Remember that the projective plane of order 3 can also be constructed as follows: Fix a point p in the affine space of order 3. Let the points and the lines of the projective plane be the lines and the subplanes containing p, respectively. Just as the affine plane of order 3 can be modelled in a natural way on a 3×3 grid, it is possible to model the affine space of order 3 on a $3 \times 3 \times 3$ grid, which, of course, is a cube. The lines through the point in the middle of the grid are just the rotation axes of the cube, and choosing p to be this point yields exactly the lines in our model.

6.5 Extending the Affine Plane to a 5-Design

We describe the construction of the famous $5 - (12, 6, 1)$ design as an extension of the affine plane of order 3. We do this with respect to the traditional picture of this affine plane on the 3×3 grid following the description in [15].

An oval in the projective plane of order 3 contains 4 points. There are 54 ovals that are completely contained in a fixed affine plane. For the rest of this section, by oval we mean one of these 54 ovals in the traditional picture of the affine plane. Given any oval, there is exactly one oval disjoint from it. We call a pair of disjoint ovals a *star*. The following diagram shows the generators for the different ovals and stars in the affine plane.

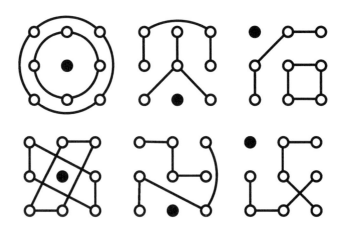

We call the ovals and stars represented by the generators in the three different rows of type A, B, and C, respectively. Given an oval, the *centered oval* associated with it is constructed as follows: Consider the star that the oval is contained in. Then there is exactly one point of the affine plane that is not contained in one of the two ovals in the star. We add this point to the oval we started with to arrive at the centered oval we are looking for.

We construct a geometry as follows: Take as the points of the geometry the 9 points of the affine plane plus three additional points A, B, and C. There are 4 different kinds of lines with 6 points contained in every single line:

- The lines of the affine plane to which the three additional points have been joined.

- The complements of lines in the affine plane with respect to the point set of the affine plane.

- Ovals of type I to which the points J and K have been joined, where $\{I, J, K\} = \{A, B, C\}$.

- Centered ovals of type I together with the point I, where I is one of A, B, or C.

There are 12 blocks each of the first and second kinds and 54 each of the third and fourth kinds.

This geometry turns out to be the classical $5 - (12, 6, 1)$ design. Deriving this design at any point gives the classical $4 - (11, 5, 1)$ design. Deriving this $4 - (11, 5, 1)$ design at a point gives the inversive plane of order 3 (see Chapter 12). Finally, deriving this inversive plane at a point brings us back to the affine plane of order 3 we started with. At any stage of the deriving process it does not matter at which points we derive, as the derived geometries are all isomorphic.

7

The Projective Plane of Order 4

The projective plane of order 4 is one of the most remarkable finite geometries. Here are some facts that make it important for us.

- The projective plane of order 4 is the only projective plane apart from the Fano plane that can be one-point extended to a 3-design. This one-point extension can be further extended, first to a $4 - (23, 7, 1)$ design and finally to the famous $5 - (24, 8, 1)$ design. See [24, p. 22] for a concise geometrical description of this extension.

- The geometry of secant lines of a fixed hyperoval turns out to be the generalized quadrangle of order $(2, 2)$. We use this fact to rebuild $PG(2, 4)$ around a two- and a three-dimensional model of this generalized quadrangle. This yields two highly symmetric models of $PG(2, 4)$. All this is something like a pictorial version of the first part of [24, Chapter 6].

- The projective plane of order 4 is the smallest projective plane of square order and therefore the smallest projective plane containing unitals. All unitals in $PG(2, 4)$ are isomorphic, and as point/line geometries they are isomorphic to the affine plane of order 3. We use this fact to rebuild $PG(2, 4)$ around a nice model of this affine plane.

- The projective plane $PG(2, 4)$ can be partitioned into three Baer subplanes. We describe such partitions in the two- and the three-dimensional models derived from the generalized quadrangle.

7.1   A Plane Model

A hyperoval in the projective plane of order 4 contains 6 points. Fix such a hyperoval and define a geometry as follows: The points of the geometry are the points of the projective plane minus the 6 points of the hyperoval. Lines are the intersections of the secants of the hyperoval with the point set of the geometry. It is easy to check that the resulting geometry is isomorphic to the generalized quadrangle of order $(2, 2)$.

We have already succeeded in extending the doily to nice pictures of the generalized quadrangles of the orders $(2, 4)$ and $(4, 2)$ and to a model of $PG(3, 2)$. Now we want to extend it to a picture of the projective plane of order 4. We just removed a set of 6 points from $PG(2, 4)$ to arrive at a representation of our favourite generalized quadrangle. Note that there are exactly 6 points of intersection in the doily that do not correspond to points of the generalized quadrangle (the solid points in the right diagram). These are exactly the points that we have to add.

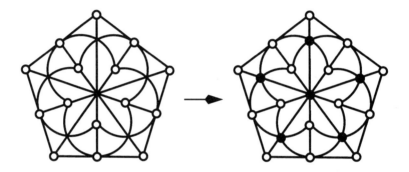

Here is a generator-only picture of $PG(2, 4)$ constructed by extending the preceding diagram. Note that the six lines that get added are the six ovoids in the generalized quadrangle (see Section 4.1).

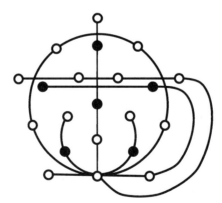

By superimposing the following two diagrams you get the complete picture (use transparencies if you actually want to do this).

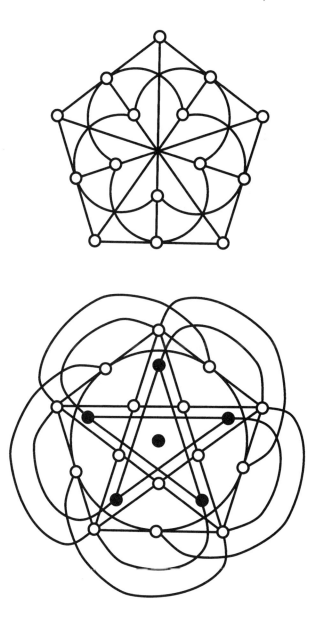

Very nice! Now let us go the other way one more time. Concentrate on the above diagram of PG(2, 4). Remove the hyperoval to discover the generalized quadrangle right in the middle. Now remove the ovoid consisting of the centers of the edges of the outer pentagon to discover the following slightly tangled up picture of the Petersen graph. We untangle it in the two following diagrams. See Section 4.2 for a picture of the generalized quadrangle with the traditional picture of the Petersen graph right in the middle.

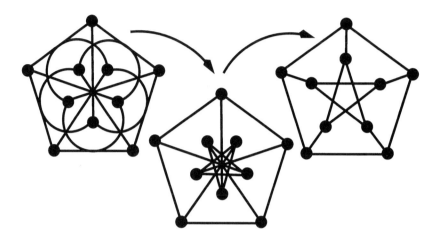

7.2   Constructing the Plane Around a Unital

Since PG(2, 4) is a projective plane of square order, it admits *unitary polarities*. The absolute points and nonabsolute lines of such a polarity form a *unital*, which turns out to be isomorphic to the affine plane of order 3. We reconstruct PG(2, 4) around the following triangular model of this affine plane that we introduced in Chapter 6.

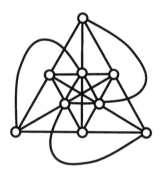

Here is how the lines of the affine plane have to be extended. Note that parallel lines in the affine plane do not get extended to lines that meet in a common point. This means that the projective completion of the affine plane cannot be embedded in PG(2, 4) around the affine plane.

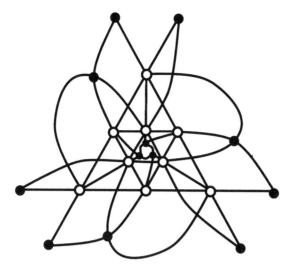

In the projective plane every line intersects the unital in either 1 or 3 points. In the above picture we see all lines that intersect the unital in exactly 3 points. Here are the generators for the 9 lines that intersect the unital in exactly one point. Note also that every point of the unital is contained in exactly one such *tangent line*.

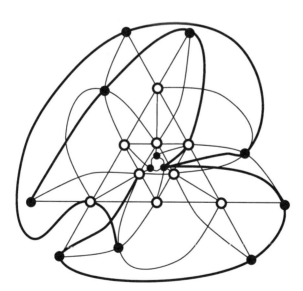

Finally, check that the set of points of any $AG(2,3)$ embedded in $PG(2,4)$ is a minimal blocking set. In fact, the set of absolute points of a unitary polarity of any classical finite projective plane is a minimal blocking set.

7.3   A partition into Three Fano Planes

The projective plane PG(2, 4) can be partitioned into 3 *Baer subplanes* that are all isomorphic to the Fano plane. The following pictures show one such partition. We use the model derived from the generalized quadrangle.

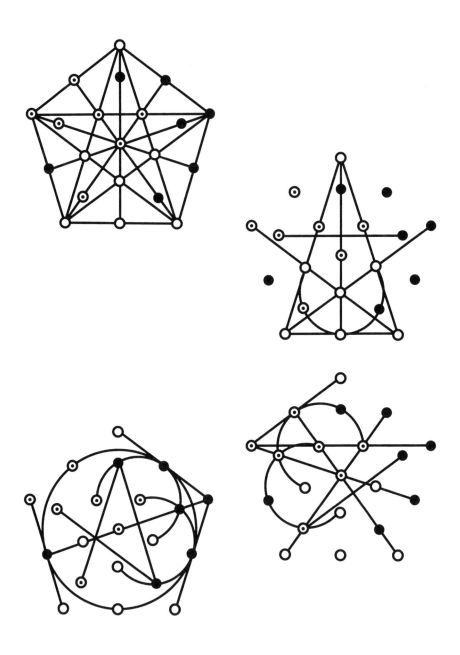

Note that the first Fano plane, consisting of the open points, is a slightly distorted traditional picture of the Fano plane. The plane consisting of the solid points is still very symmetrical; the plane consisting of the dotted points is a bit of a mess. Note, also, that every line of $PG(2,4)$ contains a line of one of the Fano planes and one point each of the other two subplanes.

We arrive at two more beautiful partitions into Fano planes in Sections 7.5 and 7.6. Note also that the set of points of any Fano plane embedded in $PG(2,4)$ is a minimal blocking set.

7.4 A Spatial Model Around a Generalized Quadrangle

As in the plane case, we extend our three-dimensional model of the generalized quadrangle to a model of the projective plane of order 4.

Here is the three-dimensional model of the generalized quadrangle we want to use. Note, again, that the edges of the tetrahedron are drawn in for reference only (see Section 4.1.2). We also introduce another frame of reference, this time in the form of an octahedron.

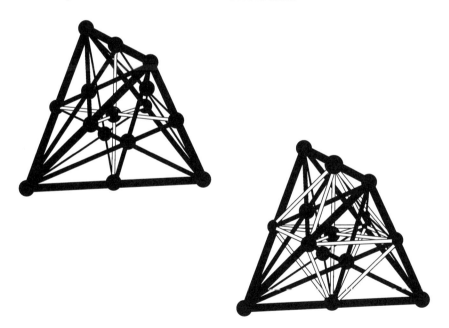

We extend our model as follows. The new model will have the full group of rotations of the tetrahedron as its automorphism group. From this we can also see immediately that the one-point stabilizer of a point in $PG(2,4)$ contains at least this group.

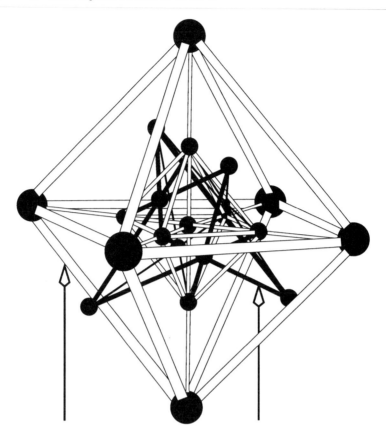

We describe what the lines in this model look like. There are 4 different kinds of lines:

- The 3 extended diagonal lines through the center of the model.

- The vertex sets of 2 tetrahedrons together with the center of the model. The first such tetrahedron is the tetrahedron of reference. The second one is the tetrahedron formed by the centers of the faces of the tetrahedron of reference. Remember that these two tetrahedrons are two of the ovoids of the generalized quadrangle (see Section 4.1.2).

- The 4 "spikes." These four spikes are the remaining four ovoids in the generalized quadrangle (see, again, Section 4.1.2).

- The 12 skew lines made up of two parts each. In the above picture the arrows point at the two parts of such a composite line. The other 11 can be constructed by rotating the model.

7.5   Another Partition into Fano Planes

The next picture gives another partition of the point set of PG(2, 4) into the point sets of three Baer subplanes. The points of the first Fano plane are the vertices of the octahedron of reference plus the center of the model. The lines in this Fano plane are the intersections of the lines of the first and third types, that is, the 3 long diagonals plus the 4 spikes with these 7 points. The resulting three-dimensional model of the Fano plane is the first model that we considered in Section 1.10.

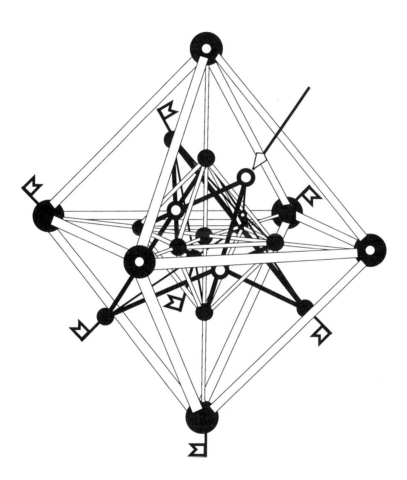

The arrangement of the other two models of the Fano plane within the larger model is still highly symmetrical. The set of 7 dotted points in the diagram above is the point set for one of the Fano planes. The remaining 7 flagged points are the points of the third Fano plane in our partition. Rotations of the model around the axis whose direction is indicated by the arrow translate into automorphisms of all three Fano planes in our partition.

A third partition into Fano planes is constructed in the next section.

7.6 Singer Diagram

Based on the difference set $\{0, 1, 6, 8, 18\}$ of order 21, which encodes all the information about $PG(2, 4)$ (see [3, p. 150]), we draw a partial Singer diagram as described in Section 1.8.

A partition of the projective plane into three Fano planes can also easily be constructed in this model of the plane. Just label the points in the diagram from 0 to 20 in the clockwise direction. The points of the first Fano plane in the partition are the 7 solid points whose labels are divisible by 3. The point sets of the second and third Fano planes are the sets of dotted and open points, respectively, that is, the points whose labels are equal to 1 and 2 mod 3, respectively.

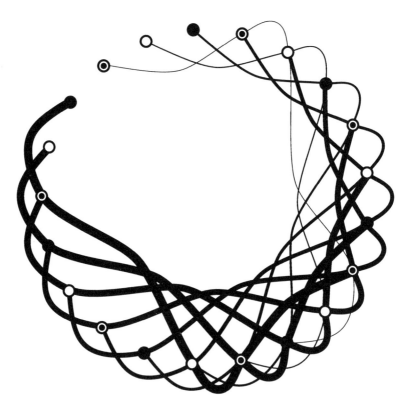

8

The Projective Plane of Order 5

8.1 Beutelspacher's Model

A detailed discussion of the following model can be found in [12]. In this paper Beutelspacher describes the projective plane of order 5 with respect to the 6 points of an oval in this plane in a manner that is very reminiscent of the way the generalized quadrangle of order $(2, 2)$ is constructed over a set of of 6 elements in terms of synthemes and duads.

The following diagram shows a generator-only model of the projective plane. Note that the line in the generator-only model below that looks like a starfish is supposed to contain the point in the center of the diagram. The six points in the middle of the diagram form an oval. The 15 points whose labels do have "two isolated points" are the exterior points of the oval. The 10 points whose labels do not have "isolated points" are the interior points of the oval. The "double" lines generate the 15 secants of the oval, the thick lines the 6 tangents, and the thin lines the 10 exterior lines of the oval. All the information about the lines is built into the labels, which all have something to do with the complete graph on the six points in the middle of the diagram. Note that the label of an exterior point consists of two edges and two isolated vertices of the complete graph and that the label of an interior point is a 1-factorization of the complete graph.

The secants correspond to the edges of the graph. The secant associated with an edge contains the two points contained in the edge plus the points whose labels contain the edge. The tangents correspond to the vertices of the graph. The tangent associated with a vertex contains the vertex itself plus all the exterior points whose labels contain the vertex as an isolated point. The exterior lines correspond to the interior points. The exterior line associated with an interior point p contains all those interior points whose labels have exactly one edge in common with the label of the point p. It also contains an exterior point if the edge connecting the two isolated vertices in its label are contained in the label of p.

The correspondence between points and lines, as we just described it, also defines a polarity of the projective plane. The absolute points and absolute lines of this polarity are the points and tangents of the oval in the middle of the diagram.

Similar constructions are possible for pictures of other classical projective planes of odd order with respect to ovals contained in them.

Here is another example: A generator-only model of the projective plane of order 3 built around an oval. Again, the line in the generator-only model pictured below that encloses the whole diagram is supposed to contain the point in the center of the diagram. The interior and exterior points and the secants and exterior lines are split up in very much the same way as above.

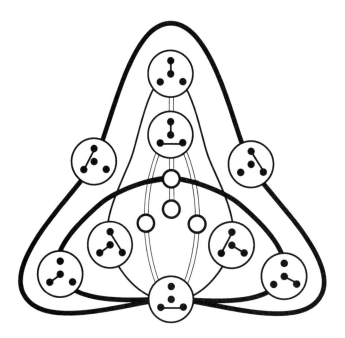

The classical projective planes of even order are better built around hyperovals. Here, for example, is a meaningful labelling of the traditional model of the Fano plane with respect to a hyperoval.

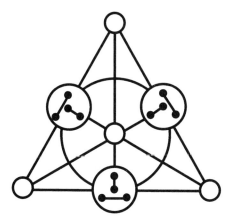

Try to use other symmetrical arrangements of 4 points in the plane and in space to come up with meaningful labellings of the different models for the Fano plane that we considered in the first chapter. Note that all our models for the projective plane of order 4 are built around hyperovals.

8.2   A Spatial Model on the Dodecahedron

The following model lives on the dodecahedron. For a detailed discussion of this model see [65].

The points of the plane are the 31 rotation axes of the dodecahedron, or equivalently, the 31 pairs of points on the dodecahedron that correspond to the pairs of points of intersections of the axes with the surface of the dodecahedron. The following pictures show the three essentially different lines in this model consisting of six axes/pairs each. The first line stands for 15 lines in the model, the second one for 6, and the last one for 10.

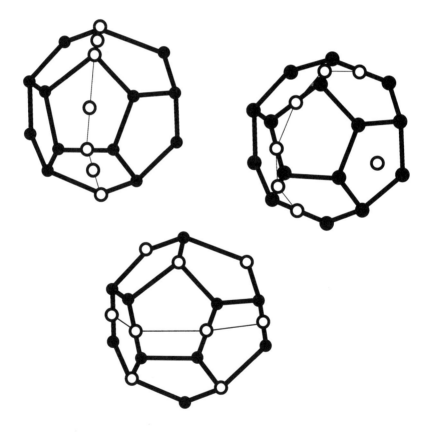

This model has counterparts for the projective planes of orders 2 and 3 on the tetrahedron and cube, respectively (see Sections 1.10 and 6.4).

The icosahedron, too, has 31 rotation axes and a model of this projective plane can also be defined in terms of these rotation axes. Just remember that the icosahedron can be inscribed into the dodecahedron such that the vertices of the icosahedron are the centers of the faces of the dodecahedron. Inscribed in this way, the rotation axes of the icosahedron and the dodecahedron coincide.

A closer look at the different kinds of points and lines in the model on the dodecahedron yields the following neat interpretation:

- The 6 axes through the centers of the faces form an oval.

- The 10 axes through the vertices and the 15 axes through the centers of the edges are the 10 interior and the 15 exterior points of the oval.

- The lines of the first, second, and third kinds correspond to the secants, the tangents, and the exterior lines of the oval.

8.3 The Desargues Configuration Revisited

An Embedding in the Projective Plane of Order 5

The set of 10 interior points together with the set of 10 exterior lines of an oval in the projective plane of order 5 forms a Desargues configuration (see [29]). This immediately yields nice embeddings of this configuration right in the centers of the two models of this plane that we just considered.

The following diagram on the left shows a generator-only model of the Desargues configuration derived from the plane model. We already pointed out that by connecting all pairs of points of a Desargues configuration that are not collinear by edges, we arrive at the Petersen graph. If we do this in this particular representation, we end up with the traditional picture of the Petersen graph on the right.

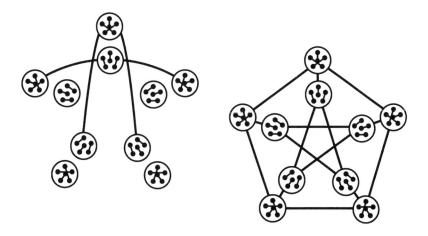

Defining the Configuration on 5 Points

Order-five symmetries of the Desargues configuration also pop up in another context: The 10 lines and 10 planes determined by 5 points in general position in real projective three-dimensional space meet any plane that is not one of the 10 distinguished ones in a Desargues configuration.

This gives rise to the following abstract description of the configuration, which is very similar to the description of the generalized quadrangle of order $(2, 2)$ in terms of synthemes and duads: Take the subsets of size two of a set with 5 elements as the points, and the subsets of three elements as the lines. This geometry is a Desargues configuration. Using this description, we can give a meaningful labelling of one of the favourite representations of the incidence graph of our configuration:

Consider the following symmetric arrangement of 5 points in space (the Schlegel diagram for the four-dimensional 5-cell).

We define the configuration in a natural way with respect to this arrangement of points: The lines are the 10 circles inscribed in the 10 triangles that you can see in this picture. The points are the 10 points in which these circles touch.

Here is a stereogram of the resulting model.

Projecting this model onto a plane gives the following picture of the Desargues configuration.

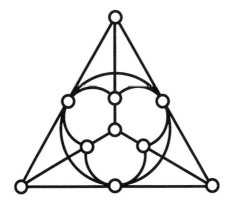

The following stereogram shows the spatial model of the Desargues configuration that we introduced on page 35. Is there an easy way to "see" that the two spatial models are models of the same configuration?

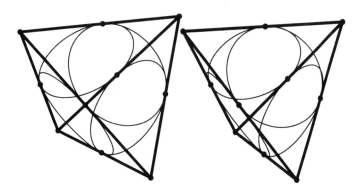

Finally, using the same setup, here is a picture of the only APC with parameters (10_3) that is not a PC (see page 35).

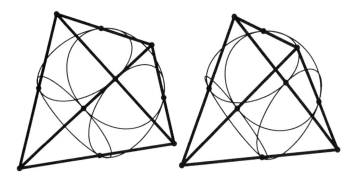

9

Stargazing in Affine Planes up to Order 8

Most diagrams in this section were originally designed by Derrick Breach and published in [15] under the title "Star Gazing in Affine Planes." In this article he presents pictures of all the affine planes up to order 8 and uses them to construct models for the classical inversive planes of the respective orders. He calls his pictures star diagrams.

Given a point in the Euclidean plane, we can partition the plane into concentric circles around this point plus the set that consists only of the point itself. Similarly, we can partition any classical affine plane into a set of ovals plus a set consisting only of one point. Breach's picture of an affine plane of order n is modelled onto a regular $(n+1)$-gon with one point of the affine plane in the center of the picture and a set of ovals partitioning the rest of the point set arranged on concentric circles. In this way the star diagrams also form something like polar coordinate systems for the respective finite affine planes.

9.1   Star Diagrams of the Affine Planes of Orders 2 and 3

Here are the two star diagrams of the affine planes of orders 2 and 3, respectively, with the ovals drawn in as circles. We have already encountered these diagrams in Section 1.3 and Chapter 6.

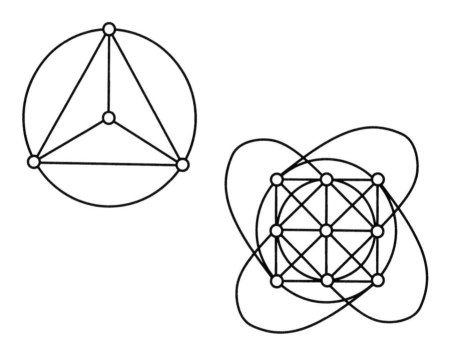

As geometries, the affine planes of odd and even orders behave very differently. Star diagrams of odd- and even-order affine planes also look somewhat different. In a star diagram of an even-order plane, any picture of one parallel class will automatically be a generator-only picture of the plane. Here are the three parallel classes in the star diagram of AG(2, 2), every one of which also is a generator-only picture.

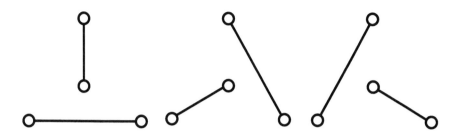

On the other hand, in diagrams of odd-order planes we have to combine two parallel classes that look quite different to arrive at a full set of generators. We will refer to two such parallel classes as a pair of generator-only pictures.

Here is such a pair for AG(2, 3).

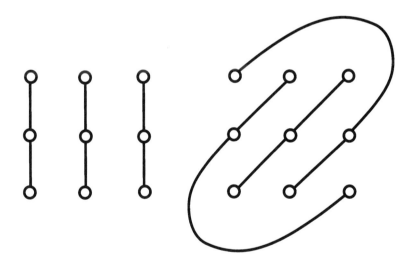

Note that there is some redundancy in providing both parallel classes. Actually, if n is the order of the plane under consideration, then $(n + 1)/2$ elements in each parallel class suffice to add up to a full set of generators.

Finally, note that in all star diagrams lines in a parallel class usually consist of lines that are made up of parallel segments. In the even case, however, every parallel class also contains one line that runs "perpendicular" to the rest of the lines in the parallel class.

9.2   Star Diagram of the Affine Plane of Order 4

The following pictures show the full star diagram, a generator-only version
of this diagram (which is also a parallel class), and the distinguished set of
three ovals. See also Section 3.3 for some additional remarks.

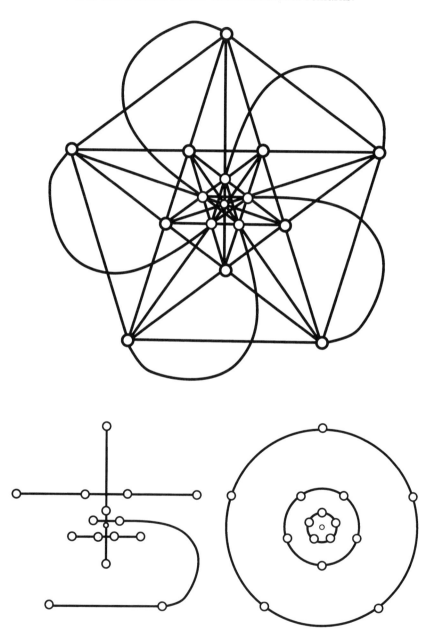

9.3 Star Diagram of the Affine Plane of Order 5

We restrict ourselves to showing the straight line diagram that underlies the star diagram together with generating parallel classes and the distinguished set of ovals.

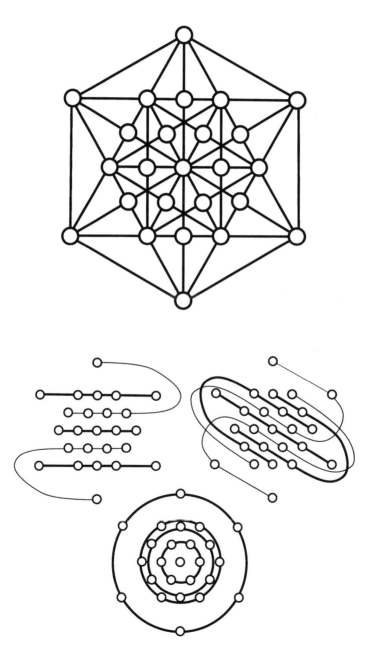

9.4   Star Diagram of the Affine Plane of Order 7

In the star diagram for the affine plane of order 7 we need only "halves" of
two parallel classes to generate all lines of the plane.

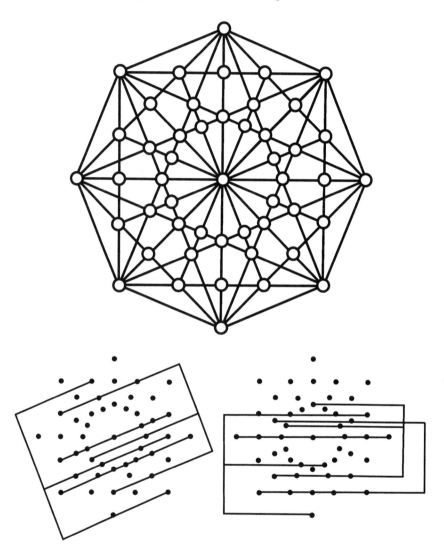

9.4.1   The Pascal Configuration in a Conic

In the planes up to order 7 all ovals turn out to be conics. Here are two
pictures of the *Pascal configuration* inscribed in a conic, the first one in a
circle in the Euclidean plane and the second one in one of the distinguished

ovals in our finite affine plane. Remember that conics in classical projective planes are characterized by the fact that all possible inscribed Pascal configurations "close." For the Pascal configuration to close just means that if one inscribes the solid part of the configuration in a conic, then the three dotted points of intersection are collinear.

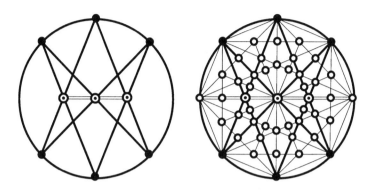

9.5   Star Diagram of the Affine Plane of Order 8

Here is the last of Breach's star diagrams together with a set of generator lines.

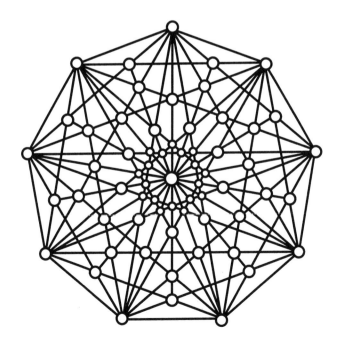

As for all star diagrams for planes of even order, all parallel classes look the same. Still, drawing all eight lines of such a parallel class leads to a very crowded situation. We therefore split this parallel class up into two parts containing four lines each.

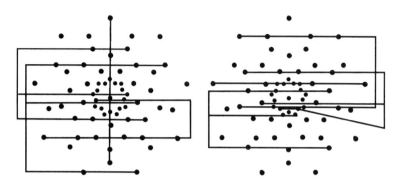

9.5.1   The Fano Configuration

Every classical projective plane of even order contains Fano subplanes. In fact, these classical planes are characterized among the other finite projective planes by the fact that all possible inscribed *Fano configurations* "close." For a Fano configuration to close just means that if one draws the solid part of the following Fano configuration, then the three dotted points of intersection are collinear (see also Section 3.1.1).

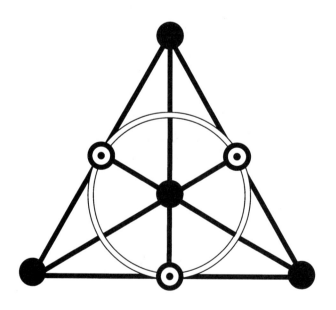

The following picture shows a Fano configuration in the projective plane of order 8. Note that the thick double circle/9-gon that completes the configuration is the line at infinity of the above affine plane.

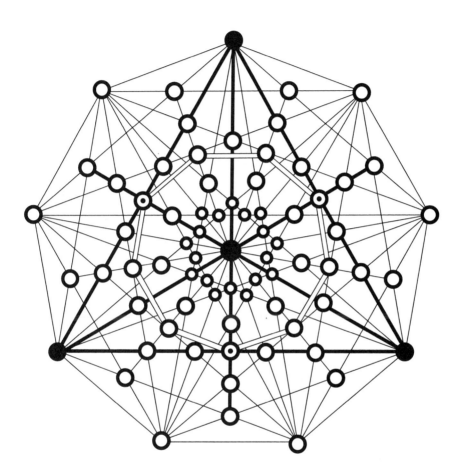

9.5.2 An Oval That Is Not a Conic

In the projective plane of order 8 there are ovals that are not conics. These ovals are constructed by replacing some point of a conic by the nucleus of the conic, that is, the common point of intersection of all the tangents of the conic. We call an oval constructed in this way a *pointed conic*. The distinguished ovals in our diagram are all conics. Their common nucleus is the center of the diagram. The following diagram shows a Pascal configuration in one of the pointed conics in the plane (the set of points contained in the simply closed curve in the diagram). Note that the large dotted point of intersection is not contained in the line connecting the other two dotted points of intersection. This means that the configuration does not close, and we can be sure that our pointed conic is not a conic.

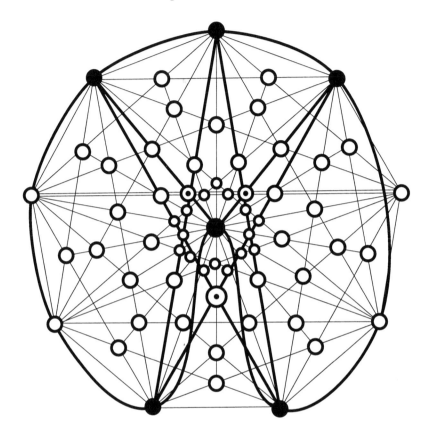

10
Biplanes

According to the *Encyclopedia Britannica*, a biplane is "an airplane with two wings, one above the other," like this one.

What the encyclopedia does not know is that *biplanes* lead a second life as geometries that satisfy the following two axioms:

Axioms for biplanes

(B1) Two distinct points are contained in two distinct lines.
(B2) Two distinct lines intersect in two distinct points.

The finite biplanes are just the symmetric $2-(v, k, \lambda)$ designs with $\lambda = 2$. In fact, they are exactly the symmetric $2-(1+k(k-1)/2, k, 2)$ designs. We call $k-2$ the *order* of a biplane. Today, to our knowledge, only 17 different finite biplanes are known, and the corresponding orders are 1, 2, 3, 4, 7, 9, and 11. It is known that there are no biplanes of orders 5, 6, and 8. There are unique biplanes of orders 1, 2, and 3, and there are exactly 3 different biplanes of order 4. Most of the mathematics behind the pictures in this section can be found in [57]. See also [45] and [91] for a complete list of the known examples of biplanes.

Here are two pictures of the unique (trivial) biplane of order 1.

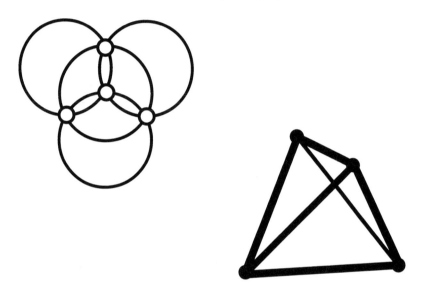

In the model on the tetrahedron, take as the points and lines the vertices and faces of the tetrahedron, respectively.

10.1   The Biplane of Order 2

The biplane of order 2 is the complement of the Fano plane; that is, its point set contains the same points as the Fano plane, and its lines are the complements of the lines of the Fano plane with respect to the point set.

A Plane Model

Here is a generator-only picture that illustrates how to go from (the traditional picture) of the Fano plane to its complement. Note that the point in the center of the diagram is supposed to be part of the line that is represented by the largest circle. The generator-only picture is followed by the full picture of the geometry.

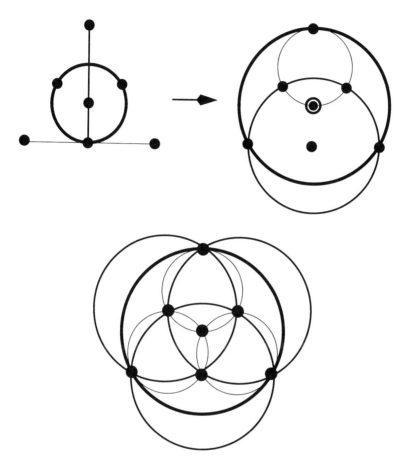

Here are the different line pencils in this biplane. Note that apart from the open carrier of a pencil, every point of the plane is contained in exactly two lines of the pencil. This also implies that our geometry really satisfies Axiom B1.

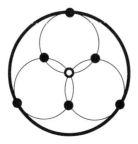

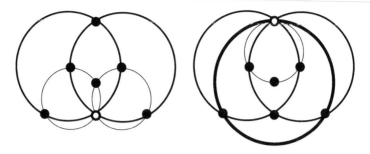

Hussain Graphs

Associated with any fixed line l of a biplane is a set of graphs, the so-called *Hussain graphs*, which are constructed as follows: There will be one graph for every point not on l. Let p be such a point. Its associated graph has the points of the line l as vertices, and two vertices are connected by an edge if the second line through the two points (different from l) also contains the point p. In our example, let the line l be the distinguished circle centered in the middle of the diagram. There are three points not on the distinguished line, corresponding to the following 3 graphs.

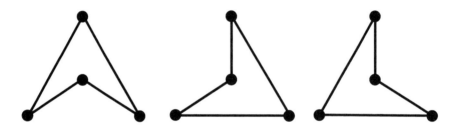

Any two graphs in a set of Hussain graphs have exactly two edges in common. Furthermore, every vertex in such a graph is contained in exactly two edges. This means that every such graph is partitioned into circles. On the other hand, given a set of n points together with a set of m graphs with these two properties, such that $n + m$ is the number of points in a biplane of order $n - 2$, then a biplane of order $n - 2$ can be constructed as follows: The Points are the n points we started out with plus the m graphs. Corresponding to every pair of points p, q there is a Line consisting of the two points plus all graphs that have an edge containing the two points as its end points. There is one more Line consisting of the n points.

In order to construct nice models of a biplane, a single line and its associated Hussain graphs can play the same role that a (hyper)oval together with a set of 1-factorizations of the (hyper)oval played in the construction of nice models for projective planes (see Chapter 8). This means that if the line has n points, then various symmetric arrangements of the n points in space will translate into different highly symmetric models of the biplane.

So, for example, we could have started by arranging 4 points in the form of a regular triangle together with its center, followed by drawing the different Hussain graphs with respect to this arrangement. Now we already have a good idea how to arrange the different kinds of points in space to arrive at a good model of the biplane.

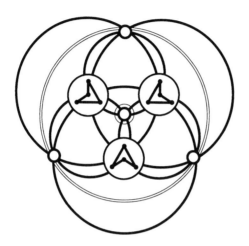

A Spatial Model

Our spatial model of the Fano plane on the tetrahedron that we introduced in Section 1.10 can of course also be turned into a nice representation of the biplane of order 2.

Here are two lines (of four open points each) that generate all the lines of the biplane (via the rotations of the tetrahedron).

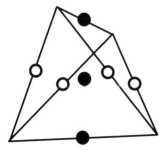 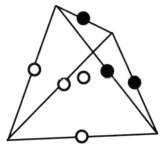

Try to come up with a meaningful labelling of this model by Hussain graphs starting with an arrangement of 4 points in a square.

Incidence Graph

The incidence graph of the biplane is the complement of the incidence graph of the Fano plane with respect to the complete bipartite graph on 14 vertices.

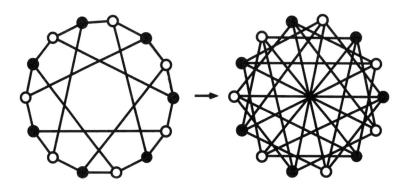

10.2   The Biplane of Order 3

A Plane Model

Here is a generator-only model of the unique biplane of order 3. It has 11 points. The labelling of this picture is via the difference set $\{1, 3, 4, 5, 9\}$. The difference set admits the number 3 as one of its *multipliers* (see [57] for a definition of multipliers). When applied to the set of labels, this multiplier has three orbits. One consists just of 0, a second one is the difference set itself, and the third one consists of the remaining 5 labels. We used this fact to arrive at this compact generator-only model of this biplane. This is followed by a picture of the different line pencils in this plane.

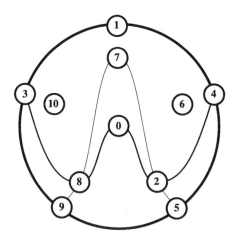

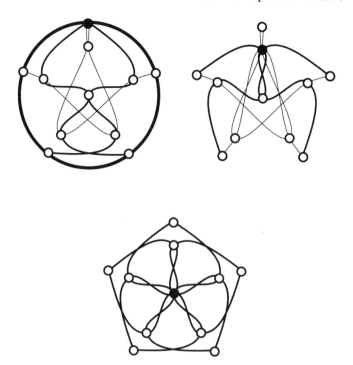

We construct the set of Hussain graphs associated with the "circle." We arrive at the following essentially different graphs. The first one corresponds to the point 0 in the middle of the diagram, the second one to the point 7. The remaining ones are the four rotated versions of the second graph.

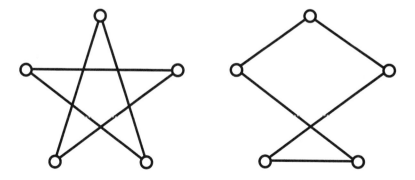

We arrive at the following meaningful labelling of the above picture of the biplane.

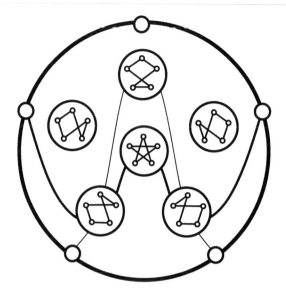

A Spatial Model

Consider the set of five points consisting of the vertices and the center of the tetrahedron. Then a set of 6 Hussain graphs is generated by the picture on the right. Note that reflecting the graph through the dashed axis leaves the graph unchanged. The axis passes through centers of opposite edges, one part of the graph, the other not. Let us call the center of the edge that is not part of the graph the "point associated with the graph."

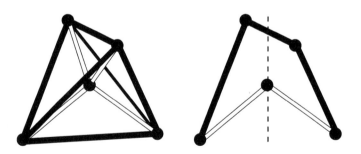

Now it is clear how to model the biplane onto the tetrahedron: Take as points the vertices, the center, and the centers of the edges of the tetrahedron, where a center of an edge is labelled with the graph it is associated with. With this labelling in place, we arrive at the following three generators for lines in the biplane. The first one stands only for itself, the second one for 4 lines, and the third one for 6 lines.

Incidence Graph

The following picture of the incidence graph of our biplane can be found in [25, p. 651].

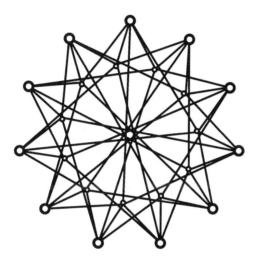

10.3   The Three Biplanes of Order 4

We already mentioned that there are exactly 3 different biplanes of order 4. Every such biplane has 16 points.

10.3.1   A First Biplane of Order 4

We first present four models of the most symmetric biplane of order 4.

The Model on the 4 × 4 Grid

First we consider the following grid. The points of the grid will be the points of the biplane. There will be one line associated with any point p (the dotted point below) of the grid. It consists of all points of the two lines in the grid through the point except the point itself.

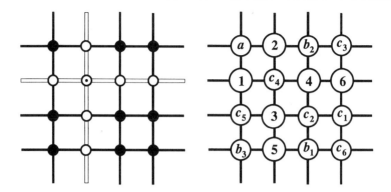

It helps to think of opposite ends of lines in the grid as being identified. Then the grid becomes a grid on the torus, and none of the points has a distinguishable position any longer. This implies that the automorphism group of this biplane acts transitively on the points of the biplane. Also, exchanging a point with its associated line defines a polarity of the biplane that does not have any absolute points or lines.

We label the points of our distinguished line from 1 to 6 as above. We arrive at the following three essentially different Hussain graphs with respect to the distinguished line corresponding to points a, b_1, and c_1. The Hussain graphs corresponding to b_2 and b_3 are the graph for b_1 rotated 120 and 240 degrees in the clockwise direction. Similarly, the graphs for c_2, c_3, c_4, c_5, and c_6 result from the graph for c_1 by successively rotating through 60 degrees.

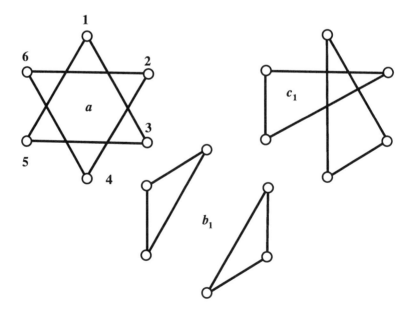

Parallelisms of Triangles

The above set of Hussain graphs has the following remarkable properties:

- Every graph partitions the distinguished set of points into triangles.

- Every triangle whose vertices are among the distinguished points is contained in exactly one of the graphs.

We call a collection of graphs on a (distinguished) set of points that has the two properties above a *parallelism of triangles* (see [22, Chapter 5]). A parallelism of triangles on $3k$ points automatically contains the right number of graphs necessary for the the construction of a biplane of order $3k - 2$. Still, a parallelism like this is not necessarily a set of Hussain graphs. Here is an example of a parallelism of triangles on 9 points generated by the following 4 graphs that is not a set of Hussain graphs.

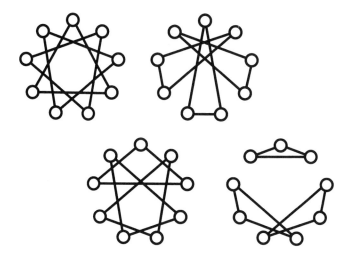

In fact, we know only two nontrivial distinct parallelisms of triangles that correspond to biplanes: the one on 6 points above and another one on a set of 9 points, which we describe in Section 10.4.

Biplanes that correspond to a parallelism of triangles have lots of sub geometries that are biplanes of order 1. These are constructed as follows: Let l be the line of the biplane that does not contain any of the graphs as points and let p be any point on this line and g any of the graphs, that is, a point not on the line l. The two lines through p and g intersect the line l in two points q and r different from p. Finally, the second line connecting q and r also contains the graph g as one of its points. This means that the graph plus the three points p, q, and r span a subbiplane of order 1.

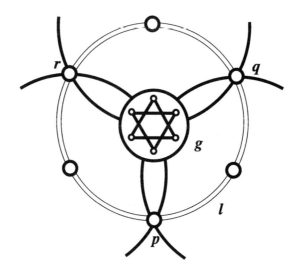

The Model on the Shrikhande Graph

Here is another very appealing picture of our biplane on the torus. Consider the two sides of the following graph, identified as indicated by the arrows, and let the points of the geometry be the vertices of the graph. Associated with every point is a line that consists of the six neighbors of the point in the graph, that is, the hexagon surrounding the point.

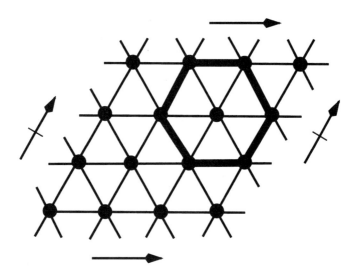

We note that this graph is the so-called *Shrikhande graph*.
Here is the graph labelled with the appropriate Hussain graphs.

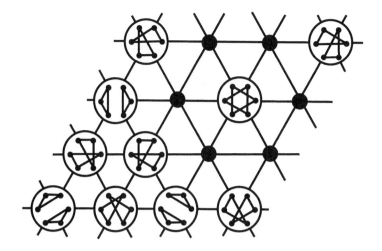

A Model on the Clebsch Graph

The points in this final model are the points of the *Clebsch graph*. Associated with every point is a line consisting of the point itself and the five points adjacent to it.

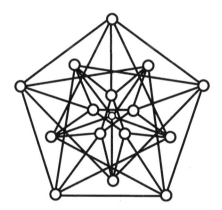

A graph Γ is called *strongly regular* with parameters (n, k, λ, μ) if it satisfies the following conditions:

- Γ has n vertices.

- Γ is regular of valency k; that is, every vertex is adjacent to exactly k other vertices.

- Any two adjacent vertices have exactly λ common neighbors.

- Any two nonadjacent vertices have exactly μ common neighbors.

Any complete graph, the Petersen graph, and the two graphs above are examples of strongly regular graphs. They illustrate the close relationship between these kinds of graphs and $2-(v,k,\lambda)$ designs that admit polarities all of whose or none of whose points are absolute. In our two models a polarity like this is defined by exchanging points with their associated lines. Given a 2-design together with a polarity like this, it is possible to construct a strongly regular graph by letting the points of the graph be the points of the design and letting two points x and y be adjacent if and only if x is contained in the image of y under the polarity. Note that if we apply this construction to the above biplane/polarity pairs, we end up constructing the graphs we started out with. The trivial $2-(4,3,2)$ biplane and one of the $2-(56,11,2)$ biplanes also admit suitable polarities that lead to interesting graphs. The strongly regular graph associated with the trivial $2-(4,3,2)$ biplane is the complete graph on 4 vertices. See [24, p. 42] for more details.

A Model on the Dodecahedron

Using the abstract definition of the set of Hussain graphs corresponding to this biplane in terms of the parallelism of triangles on a set of six elements, we proceed to model the biplane on the dodecahedron: Take as the distinguished 6 points the set of 6 axes through the centers of the faces of the dodecahedron. Then all Hussain graphs look like the generator on the left.

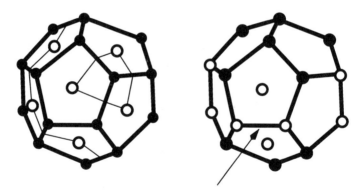

Now we can take as the points of the biplane the 6 axes through the centers of the faces and the 10 axes through the vertices. One of the lines consists of the 6 axes through the centers of the faces. The other lines are generated by the lines consisting of the axes through the open points in the diagram on the right. The arrow points at an edge. The rotation axis of the dodecahedron through the center of this edge is a symmetry axis for this second kind of line. There are 15 such rotation axes corresponding to the 15 lines of the second kind.

It turns out that all lines in this model of the biplane are ovals in the model of the projective plane of order 5 on the dodecahedron that we considered before. See [19] for more information about this embedding.

A Model on the Tesseract

This biplane of order 4 can also be modelled onto the tesseract, that is, the four-dimensional hypercube. The following is a well-known representation of this four-dimensional "regular solid" in three-dimensional space.

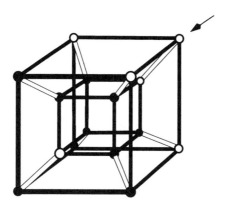

The points of the biplane are the vertices of the hypercube. There is one line associated with any point p (the arrow in the picture points at this point). This line consists of the points of distance 0 (the point itself), 1 and 4 from the distinguished point. This gives a total of 6 points highlighted in the picture. As in our first model, exchanging a point with its associated line defines a polarity of the biplane. There is a difference, though, in that all points and lines of the biplane are absolute.

We will come back to this picture in Section 11.1.

10.3.2 A Second Biplane of Order 4

We are going to present two models for this biplane.

A Model on the 4 × 4 Grid

In a first model of the biplane we again consider the 4 × 4 grid. The picture on the left shows four "diagonals" that partition the 16 points of the grid. Every diagonal contains 4 points. The points of the grid will be the points of the biplane. There will be one line associated with any point of the grid. It consists of the six points that are contained neither in the two lines through the point nor in the diagonal through the point. The picture on the right shows the line associated with the point in the upper left corner of the grid.

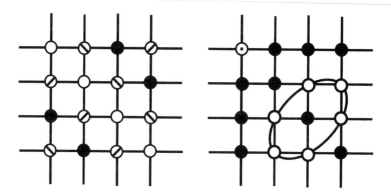

A Model on the Hexagon

The following picture is a generator-only picture of this second-most-symmetric biplane of order 4. It has been reconstructed from the set of Hussain graphs listed in [57, Fig. 3.5]. Note that antipodal points on the outer circle of the diagram are to be thought of as identified. Note also that the line that is point-symmetric with respect to the center of the model does not contain the center.

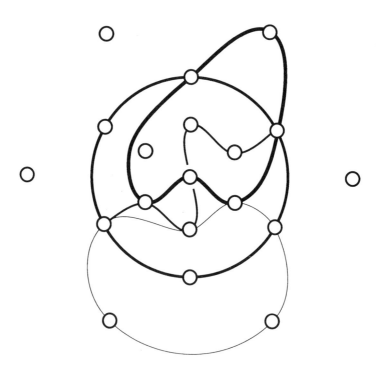

Because of the rotational symmetry, we get three essentially distinct Hussain graphs associated with the "circle."

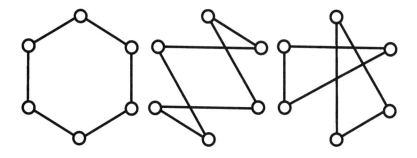

Can you show that the two models in this section really describe the same biplane?

10.3.3 A Third Biplane of Order 4

We have not been able to find a reasonably appealing picture of the third biplane of order 4. Here is only one possible set of Hussain graphs for this biplane. For more information about this biplane see [57].

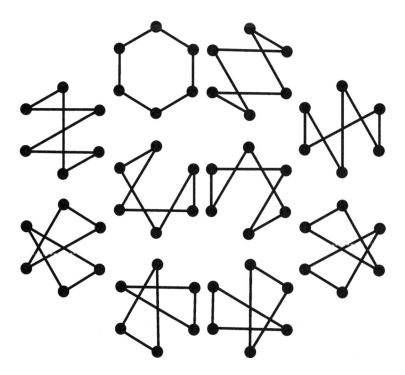

10.4   Two Biplanes of Order 7

10.4.1   A First Biplane of Order 7

We exhibit just a set of generators for the Hussain graphs that correspond
to this biplane. The first generator stands only for itself, the other three for
9 graphs each. This set of Hussain graphs is also a parallelism of triangles.
Compare this parallelism to the one on page 145.

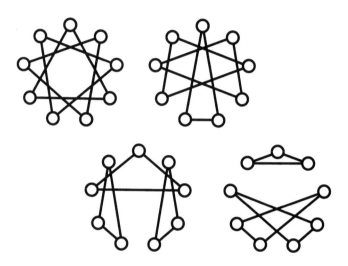

If you feel like drawing a generator-only picture of this biplane, start with
4 nested regular 9-gons plus a point in the common center of these 9-gons,
let the 9 distinguished points of the biplane be the 9 vertices of the outer
9-gon, label the point in the middle with the first of the graphs, and label
the vertices of the remaining three 9-gons with the graphs generated by the
second, third, and fourth graphs in the obvious way. Finish by drawing in
the the 5 generators for the different blocks: one that is a circle containing
the 9 distinguished points and 4 more that correspond to the 4 essentially
different diameters of the regular 9-gon.

10.4.2   A Second Biplane of Order 7

All the information about our second biplane is given by the difference
set $\{1, 7, 9, 10, 12, 16, 26, 33, 34\}$ of order 37 (see [57, Example 2.5]). This
difference set admits the number 7 as one of its multipliers. When applied
to the difference set, this multiplier has 5 orbits. One of the orbits consists
just of the number 1; the other four orbits contain 9 numbers each. Use
this observation to draw a generator-only picture of this biplane. Start by
figuring out how exactly we constructed the picture of the biplane of order 3
on page 140.

10.5   A Biplane of Order 9

The projective plane of order 4 contains 168 hyperovals. Fix one of these hyperovals. We define a geometry as follows: Let the Points of the geometry be the 56 hyperovals that have an even number of points in common with the fixed hyperoval. Associate with every Point/hyperoval a Line of the geometry consisting of the Point/hyperoval itself and all Points/hyperovals that do not have any points in common with the Point/hyperoval. This geometry turns out to be a biplane of order 9. See [50, p. 390].

Compare this construction to the construction of the first biplane of order 4 from ovals in the projective plane of order 5 that we described earlier. Also, we noted before that the complement of a line in the Fano plane is a hyperoval. Hence, the biplane of order 2 can also be considered as the geometry of hyperovals in the Fano plane.

10.6   Blocking Sets

In this section we give some examples of minimal blocking sets in the biplanes that we encountered in this chapter. This section is based on [5] and [6]. As a consequence of the axioms that biplanes satisfy, it is clear that in a biplane of order n every subset of a line containing $n + 1$ points is a minimal blocking set. We call these blocking sets *trivial*.

In the biplane of order 1 the minimal blocking sets are the two-point sets.

In the biplane of order 2 the nontrivial minimal blocking sets are the sets containing 4 points that are not lines.

In the biplane of order 3 the nontrivial minimal blocking sets are constructed as follows: Consider three lines that have a common point of intersection. Then the set of all points of intersection of these lines contains 4 points and forms a minimal blocking set.

Here is an example. We draw only the three lines that intersect in the solid point at the top of the diagram.

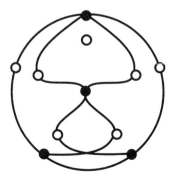

The solid points are the points of the blocking set. Convince yourself that this blocking set is minimal.

Minimal blocking sets in the biplanes of order 4 contain 4, 5, or 6 points.

Consider the grid model of our first biplane of order 4. In this model the set of four points contained in any of the verticals or horizontals in the grid form a blocking set of size 4. This means that the four blocking sets that correspond to the verticals in the grid form a partition of the plane into blocking sets. Try to construct a similar partition of our second biplane of order 4.

The next picture shows a trivial minimal blocking set of size 5 and a minimal blocking set of size 6, again in the grid model of our first biplane of order 4.

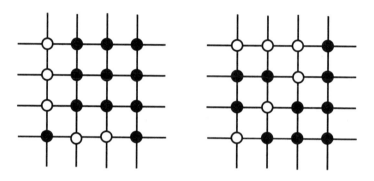

A minimal blocking set of size 6 in a biplane of order 4, as the one on the right, has some very interesting properties:

- There are exactly 10 lines that intersect the blocking set in exactly 3 points. Let us call these 10 sets of three points "Lines."

- The remaining 6 lines intersect the blocking set in exactly 1 point.

- The geometry consisting of the 6 points of the blocking set and the 10 Lines is isomorphic to the $2 - (6, 3, 2)$ design that we considered in Section 2.1.

11
Semibiplanes

Those readers who have not come across mathematical biplanes will probably think of a semibiplane as something like this.

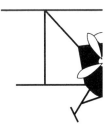

In reality, a *semibiplane* is a connected geometry that satisfies the following axioms:

Axioms for semibiplanes

(S1) Two points are contained in no or exactly two distinct lines.
(S2) Two lines intersect in no or exactly two distinct points.

Of course, semibiplanes are generalizations of biplanes, and every biplane is also a semibiplane. As with finite biplanes, in a finite semibiplane the number of lines equals the number of points, and the number of points on a line equals the number of lines through a point. In contrast to biplanes, the number of points on a line does not determine the total number of points of the geometry. We say that a semibiplane is of *order* (v, k) if v is the total number of points of the geometry and k is the number of points on a line. Infinitely many different finite semibiplanes are known to exist. For more information about semibiplanes see [57], [115], [117], [118], and [119].

In [115] a complete list of the finite semibiplanes with up to 6 points on a line is given. We list these semibiplanes in the following table.

k	v	description
3	4	biplane
4	7	biplane
	8	*hypercube*
5	11	biplane
	12	*icosahedron*
	14	biplane of order 4 doubled
	16	*hypercube*
6	16	3 biplanes
	18	3 proper semibiplanes
	22	biplane of order 5 doubled
	24	2 proper semibiplanes
	28	biplane of order 4 doubled twice
	32	hypercube

In the following we describe the semibiplanes whose descriptions in this table are given in italics.

11.1   The Semibiplanes on Hypercubes

The n-dimensional hypercube H_n is a graph whose point set coincides with $\mathbf{Z}_2^n$ and two of whose vertices are connected by an edge if and only if they differ by only one entry. The hypercube H_n has 2^n vertices. We construct a geometry whose points are the points of the hypercube. Furthermore, we associate with every point of the hypercube a line consisting of $n + 1$ points. This line contains the point itself plus all the points at distance 1 from this point on the hypercube. The resulting geometry turns out to be a semibiplane of order $(2^n, n + 1)$. We are interested only in the hypercubes of dimensions 2, 3, and 4. These are the square, the cube, and the tesseract that we already looked at in the section on biplanes of order 4.

The following set of pictures illustrates the semibiplane on the square. From the above table it is clear that this semibiplane is just the trivial biplane of order 1. The picture on the left is a generator-only diagram. The following two pictures show the different relative positions of two lines in the geometry.

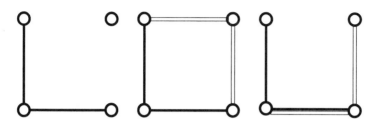

The next picture does the same for the semibiplane on the cube that is the smallest semibiplane that is not a biplane.

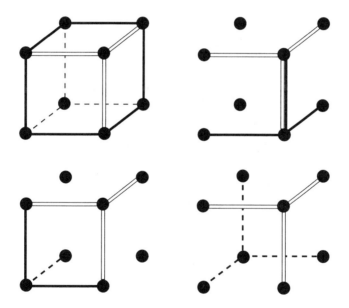

Note that the two lines in the last diagram do not intersect and that nonintersection of lines partitions the line set into three sets, each of which contains two "opposite" lines. This implies that this semibiplane is *divisible*.

We construct the Hussain graphs associated with a (any) line in this semibiplane. As in the case of biplanes, semibiplanes are determined completely by a set of Hussain graphs like this (see [116] for details).

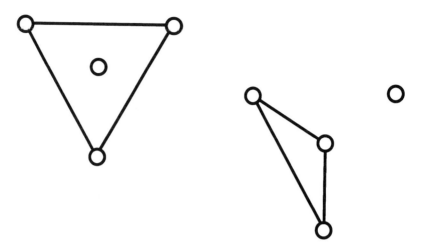

The next diagram shows a line in the semibiplane on the tesseract. Note that we can extend this semibiplane to the first biplane of order 16 that we considered in Section 10.3 by adding one point each to the lines of the semibiplane (see also [57, Exercise 7.16]).

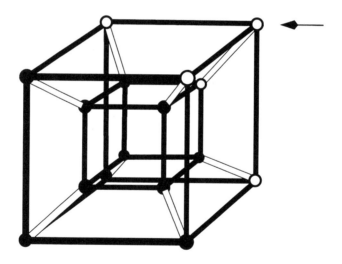

11.2   The Semibiplane of Order $(12, 5)$ on the Icosahedron

The points of this geometry are the points of the icosahedron, and the lines are the regular pentagons on the icosahedron. We found the idea for this representation in [70]. The following set of diagrams illustrates, as in the case of hypercubes, what lines look like and the different relative positions of two lines in the geometry.

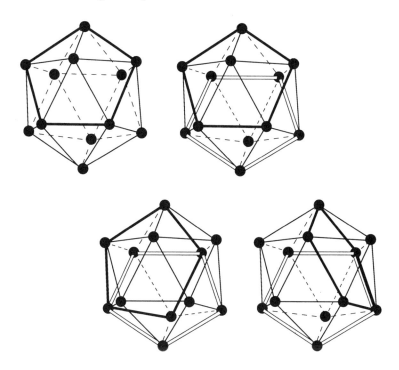

This geometry, like the last one, is divisible. Note the close similarity with the biplane of order 4 modelled onto the Shrikhande graph.

Here are the Hussain graphs associated with a (any) line of the plane.

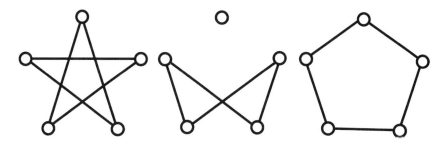

11.3   Folded Projective Planes

There is a way to construct semibiplanes from projective planes that admit an involutory automorphism. All classical projective planes admit such automorphisms, and for the projective planes of small orders some of these automorphisms correspond to reflections of our diagrams in some of their axes of symmetry.

Let P be a projective plane of order q and let π be an involutory automorphism of the plane. Then a semibiplane can be constructed as follows: Let the Points be the unordered pairs of distinct points of the projective plane that get exchanged by the involution and let the Lines be the unordered pairs of distinct lines that get exchanged by the involution.

If π is a *Baer involution* (that is, q is a square, and the fix-geometry of the involution is a Baer subplane), then the resulting semibiplane is of order $((q^2 - \sqrt{q})/2, q)$. The only projective plane of square order that we are considering in this book is $PG(2, 4)$. The semibiplane constructed from this plane via an involution like this is of order $(7, 4)$ and is therefore the unique biplane of order 2.

If q is even and π an elation, then the resulting semibiplane is of order $(q^2/2, q)$. This implies that $PG(2, 4)$ gives a semibiplane of order $(8, 4)$ that is the semibiplane modelled on the cube. Furthermore, $PG(2, 8)$ gives a semibiplane of order $(32, 8)$.

If q is odd and π a homology, then the resulting semibiplane is of order $((q^2 - 1)/2, q)$. Now, $PG(2, 3)$ gives the unique biplane of order 1, $PG(2, 5)$ the unique semibiplane of order $(12, 5)$ on the icosahedron, and $PG(2, 7)$ a semibiplane of order $(24, 7)$.

12
The Smallest Benz Planes

Benz planes comprise three types of geometries: *Möbius planes* (or *inversive planes*), *Laguerre planes,* and *Minkowski planes.* We first give descriptions of the inversive planes of orders 2 and 3 before giving a unifying definition for all three kinds of Benz planes. Standard references for Benz planes are [4] and [103].

12.1 The Smallest Inversive Planes

The *inversive planes,* or *Möbius planes,* of order n are the $3-(n^2+1, n+1, 1)$ designs. They are also characterized as those geometries all of whose derived geometries are affine planes of order n. The classical examples arise as the nontrivial plane sections of ovoids in the three-dimensional projective space of order n. In fact, if n is even, then it can be proved that all inversive planes of order n arise in this manner. The "classical" examples of inversive planes are the inversive planes associated with ovoids that are elliptic quadrics, and in the odd-order case all known inversive planes are classical.

12.1.1 The Inversive Plane of Order 2

The inversive plane of order 2 is the unique $3-(5, 3, 1)$ design, that is, the trivial geometry of all subsets of size 3 of a set with five elements. Here are two nice pictures of this geometry. We have already encountered this geometry on page 76.

A Spatial Model

The points of this model are the center and vertices of a tetrahedron, and its circles are all triangles formed of three of these points.

It is clear that this is really a model of the inversive plane of order 2. Still, it is interesting to have a look at the derived geometries at two essentially different points of the model: the center and one of the vertices. Deriving at the center yields that the 4 vertices of the tetrahedron and its six edges form one of the most symmetrical models of the affine plane of order 2. Note also that the parallel classes in this model are just the 3 pairs of opposite edges.

A Model on a Pentagon

The points of the geometry are the 5 edges of the following pentagon, and its 10 circles are generated by the two circles on the right.

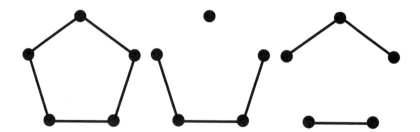

12.1.2   The Inversive Plane of Order 3

The inversive plane of order 3 is the unique $3 - (10, 4, 1)$ design. Here are two nice descriptions of this geometry.

A Spatial Model

It is again possible to model this geometry onto the tetrahedron such that all symmetries of the tetrahedron are automorphisms of the geometry. Let the points of the geometry be the 4 vertices and 6 centers of the edges of the tetrahedron. The complete set of circles of the geometry can be generated by applying all possible rotations of the tetrahedron to the following highlighted circles. The number under each circle is the number of its images under such rotations. The numbers add up to 30, the total number of circles in the inversive plane.

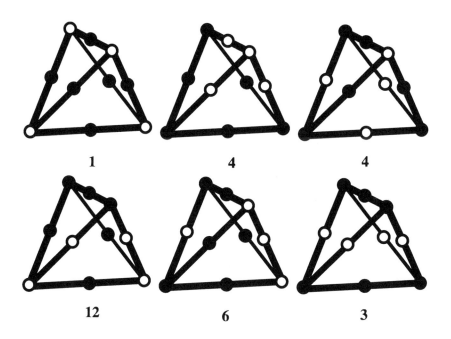

In order to verify that we are indeed dealing with a model of the inversive plane of order 3, we check that the derived geometries at one of the vertices and one of the centers of an edge of the tetrahedron are isomorphic to the affine plane of order 3.

One highly symmetric view of the point set of our model is depicted below on the left. If we derive at the point in the middle, which corresponds to one of the vertices of the tetrahedron, we get the generator-only model of the derived geometry on the right.

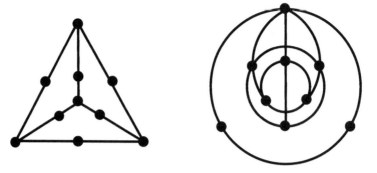

Convince yourself that this geometry is really a model of the affine plane of order 3 using the following partition of its line set into parallel classes.

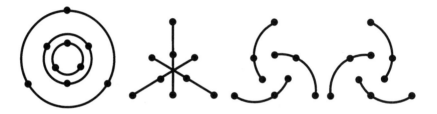

There is another highly symmetric view of the tetrahedron that we are going to use to construct the derived geometry at the center of one of the edges of the tetrahedron.

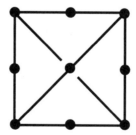

We derive at the point in the middle of the diagram that is not visible. The following four diagrams represent a partition of the line set of the derived geometry. Convince yourself, using this partition, that the derived geometry is again the affine plane of order 3.

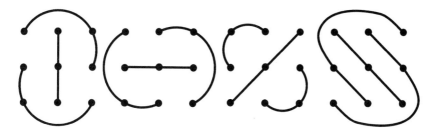

A Model on the Pentagon

Take as the points the 10 edges of the following complete graph on five vertices. The lines are the 30 subgraphs of this graph generated by the six subgraphs surrounding the complete graph.

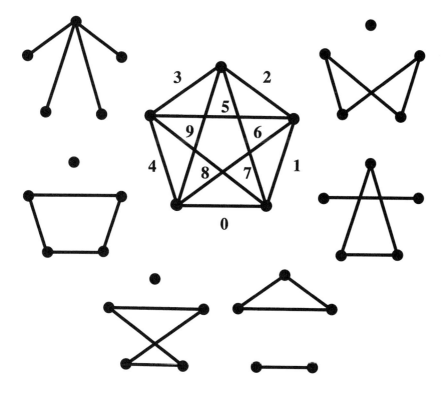

This representation is a corrected version of [24, Example 2.2, p. 30].

In order to check that this geometry is indeed the inversive plane of order 3, we only have to convince ourselves that the derived geometries at the two points that correspond to the horizontal edges, labelled 0 and 5, are isomorphic to the affine plane of order 3. With the above labelling of the 10 edges, we find that this is indeed the case. The following pictures show the derived affine planes at the points/edges 0 and 5, respectively.

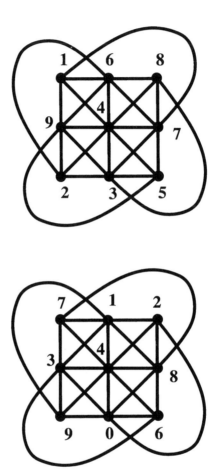

The inversive plane of order 3 can be extended even further, first to a $4 - (11, 5, 1)$ design and then, one more time, to the famous $5 - (12, 6, 1)$ design. See Section 6.5 for a rough description and [15, p. 26] and [24, p. 23] for a more detailed description of this extension process. The inversive plane of order 3 is the only inversive plane that can be extended by one further point to a 4-design.

Breach [15] gives descriptions of all inversive planes up to order 8 based on the star diagrams of affine planes that we introduced in Chapter 9.

12.2 A Unifying Definition of Benz Planes

The following definitions of Benz planes can be found in [103].

A *Benz plane* is a geometry whose point set is provided with one or two equivalence relations, called *parallelisms*. Two points are parallel if they are related by any of these equivalence relations. The equivalence classes are called *parallel classes*. A parallelism is trivial if two points are parallel with respect to it only if they coincide. Furthermore, any Benz plane satisfies the following axioms:

Axioms for Benz planes

(B1) Three pairwise nonparallel points are contained in a uniquely determined circle.

(B2) Given a point p on a circle c and a point q not parallel to p, there is a uniquely determined circle that contains both points and *touches* c, that is, intersects c only in p or coincides with c.

(B3) Parallel classes with respect to a nontrivial parallelism and circles intersect in a unique point.

(B4) Parallel classes with respect to different nontrivial parallelisms intersect in a unique point.

(B5) Each circle contains at least three points.

If a Benz plane has two different nontrivial parallelisms, it is a *Minkowski plane*. If it has only one nontrivial parallelism, it is a *Laguerre plane*. In this case Axiom B4 does not apply. If it has only the trivial parallelism, both Axioms B3 and B4 do not apply, "nonparallel" translates into "distinct," and the Benz plane is a Möbius plane. For finite Benz planes, Axiom B2 is a consequence of Axioms B1, B3, B4, and B5; every circle has a constant number $n + 1$ of points; and n is called the order of the plane. Furthermore, every single one of the parallel classes of a finite Laguerre plane of order n contains n points, and every parallel class of a Minkowski plane of order n contains $n + 1$ points.

The classical examples of Möbius, Laguerre, and Minkowski planes of order n arise as the geometries of nontrivial plane sections of an elliptic quadric, elliptic cone, and hyperbolic quadric, respectively, in three-dimensional projective space over the field with n elements.

The *complete derived geometry* at a point p of a geometry having nontrivial parallelisms is the derived geometry at this point whose line set has been complemented by all parallel classes that do not contain p. It turns out that a finite geometry having up to two nontrivial parallelisms satisfying Axioms B3–B5 is a Benz plane if and only if the complete derived geometries at all points are affine planes. The complete derived geometry of a Benz plane of order n is an affine plane of order n.

12.3   The Smallest Laguerre Planes

In this section we present models of the Laguerre planes of orders 2 and 3. Both these planes can be drawn on grids on a torus. We will represent this torus by a cylindrical grid whose top and bottom ends are to be identified in the obvious way. The rotations/translations along grid lines correspond to automorphisms of the planes. We will therefore give only a set of generating circles from which the complete set of circles can be reconstructed by rotating the torus in the two possible directions. The number below a generating circle is the number of different circles that can be generated from it in this manner.

12.3.1   The Laguerre Plane of Order 2

The verticals in the following diagrams are the parallel classes, the sets of solid points, the circles. We need only two generating circles.

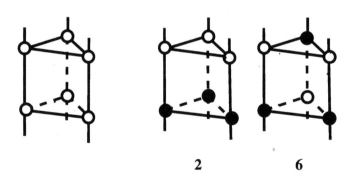

$$2 \qquad\qquad 6$$

We check that we are indeed dealing with a Laguerre plane. Since this Laguerre plane has a point-transitive automorphism group, it suffices to check that the complete derived geometry at one point is the affine plane of order 2. The following diagram shows the complete derived geometry at the highlighted point. It is indeed the affine plane of order 2. We also list the four different circles containing the point.

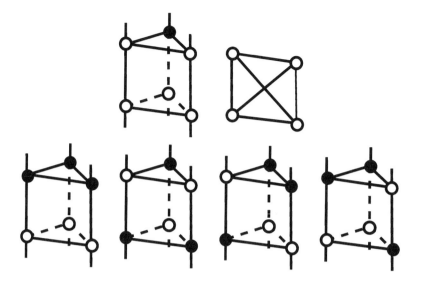

This smallest Laguerre plane also has a very appealing representation on the octahedron: Take as points of the geometry the six vertices of the octahedron, call two points parallel if they are not connected by an edge, and take as circles of the geometry the 8 faces of the octahedron.

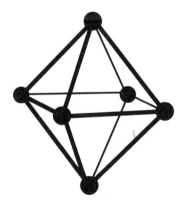

Note that this geometry on the octahedron has the property that 3 pairwise nonparallel points are contained in a unique circle and that the derived geometry at any of its points can be considered as a geometry on the tetrahedron such that any 2 nonparallel points are contained in a unique circle/line. It is possible to extend this appealing observation by constructing a geometry on the so-called 16-cell. This is, like the tesseract, one of the four-dimensional regular solids. It consists of 8 vertices and 16 cells each one of which is a tetrahedron (see [49] for details). Here is one projection of this 16-cell into three-dimensional Euclidean space. The geometry has the vertices of the 16-cell as points. Two points are called parallel if they are not connected by an edge, and the circles of the geometry are the 16 cells of the regular solid. Then any four pairwise nonparallel points are contained in exactly one circle, and the derived geometry at any of its points is the smallest Laguerre plane (on an octahedron).

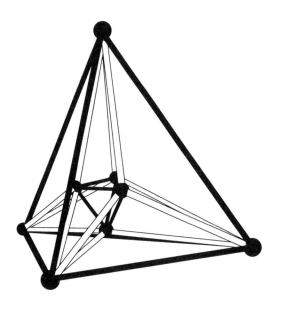

12.3.2 The Laguerre Plane of Order 3

The same remarks as in the order 2 case apply. Check for yourself that the following geometry is really the Laguerre plane of order 3 by identifying the derived geometry at some point with the affine plane of order 3.

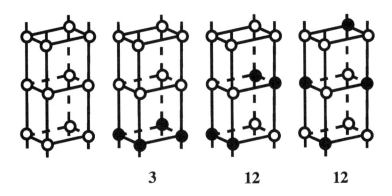

3   **12**   **12**

12.4   The Smallest Minkowski Planes

In this section we present models of the Minkowski planes of orders 2 and 3. Basically the same remarks as in the section on the smallest Laguerre planes apply. The only difference now is that there are two different kinds of parallel classes forming a grid.

12.4.1   The Minkowski Plane of Order 2

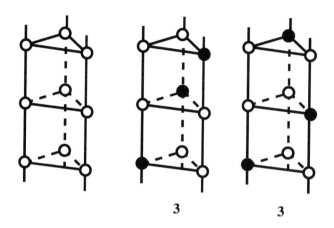

It is interesting to note that the geometry that shares the point set with this Minkowski plane and whose lines are the circles and parallel classes of the Minkowski plane happens to be the affine plane of order 3.

12.4.2   The Minkowski Plane of Order 3

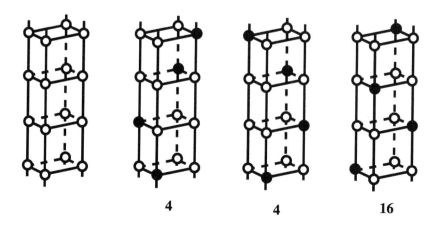

13
Generalized Polygons

A *generalized n-gon*, $n \geq 3$, is a geometry that satisfies the following axiom:

Axiom for generalized n-gons

(Po) The incidence graph of the geometry has diameter n and girth $2n$.

This axiom implies that in a generalized n-gon it is impossible to draw i-gons for $2 \leq i < n$ but that it is possible to draw n-gons. In particular, it implies that two points in a generalized n-gon are contained in at most one line, and that two lines intersect in at most one point. Any n-gon in a generalized n-gon is called an *apartment*.

In finite generalized n-gons all lines contain the same number $s + 1$ of points, and all points are contained in the same number $t + 1$ of lines. As usual, (s, t) is called the *order* of the generalized n-gon. A projective plane of order n is a generalized 3-gon, or generalized triangle, of order (n, n). The generalized 4-gons are the generalized quadrangles that we introduced in Chapter 4. Generalized 5-gons, 6-gons, etc. are also called generalized pentagons, hexagons, etc. The dual of a generalized n-gon of order (s, t) is a generalized n-gon of order (t, s). The ordinary n-gon is the unique generalized n-gon of order $(1, 1)$.

From now on all generalized n-gons are supposed to be finite. Unless $s = t = 1$, the only possible values for n are 3, 4, 6, 8, or 12. The value $n = 12$ is possible only if either s or t equals 1. The incidence graphs of generalized triangles (such as the projective planes), quadrangles, and hexagons of order (q, q) are precisely the generalized hexagons, octagons, and 12-gons of order $(1, q)$. See [21, Chapter 9] for more information about generalized n-gons.

13.1   The Generalized Hexagon of Order $(1, 2)$

This hexagon is the incidence graph of the Fano plane. We already showed how to draw nice pictures of this and other incidence graphs of projective planes in the chapter on the Fano plane.

Here it is again surrounded by 4 pictures showing the different ways an apartment in this geometry can be situated in the diagram.

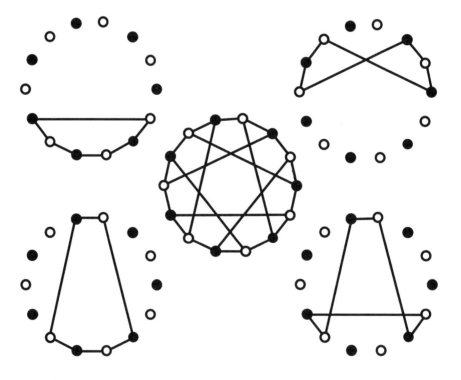

The line set of this generalized hexagon, just like the line set of any other generalized hexagon of order $(1, n)$, admits a partition into spreads.

One the other hand, all these hexagons also contain ovoids. The set of points that correspond to points of the Fano plane is an ovoid. How many other ovoids are there?

13.2   The Generalized Hexagon of Order (1, 3)

This generalized hexagon is the incidence graph of the projective plane of order 3. Here it is again followed by a list of different ways apartments are situated in this geometry. This list is not complete. See how many other apartments you can find in this diagram.

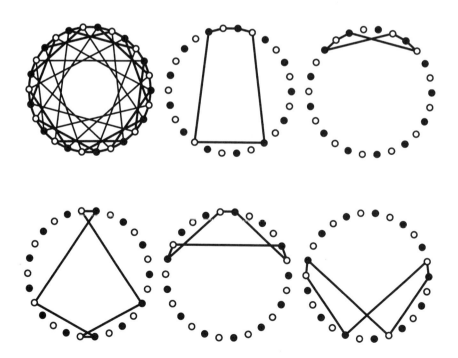

13.3 The Two Generalized Hexagons of Order $(2, 2)$

There are two generalized hexagons of order $(2, 2)$, one the dual of the other. Andreas Schroth [94] has managed to draw some nice pictures of these geometries based on automorphisms of order 7 of these hexagons. Here is the first one.

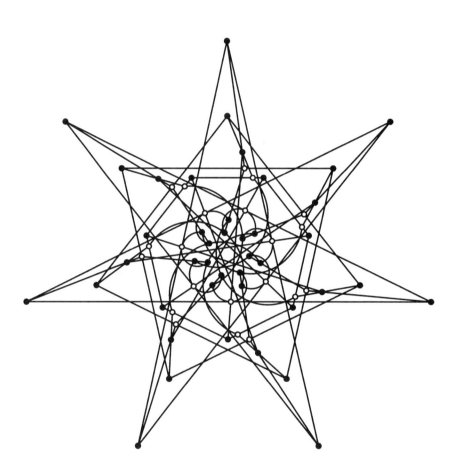

Note that the set of open points forms an ovoid in this generalized hexagon. This means, just as in the case of generalized quadrangles, that every line in the geometry contains exactly one of the points of this distinguished set. This generalized hexagon does not contain spreads. All kinds of other geometries such as the Fano plane and the generalized quadrangles of orders $(2,2)$ and $(2,4)$ can be constructed from this generalized hexagon (see [95] and [74, p. 42]).

An elementary construction of this first hexagon is given in [114]: Take as points the points of $PG(2,9)$ minus the absolute points of a unitary polarity, that is, the points of a unital, and as lines the self-polar triangles of the polarity. Note that this construction is very similar to the construction of the generalized quadrangle of order $(2,2)$ from $PG(2,4)$ that we used in Chapter 7. This generalized hexagon is also known as the "$G_2(2)$ hexagon" since the Chevalley group $G_2(2)$ is its automorphism group.

The following picture shows the dual of the generalized hexagon depicted above.

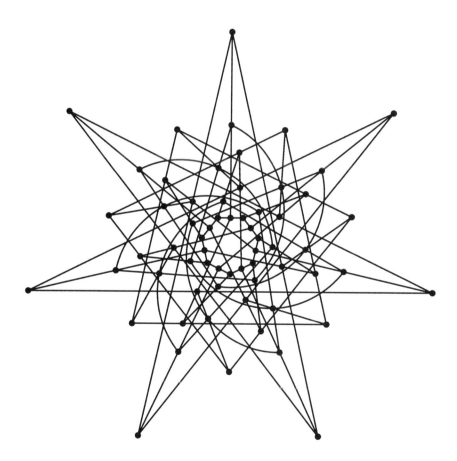

As the dual of the first hexagon, the second generalized hexagon does not contain ovoids. What it does contain are spreads such as the following.

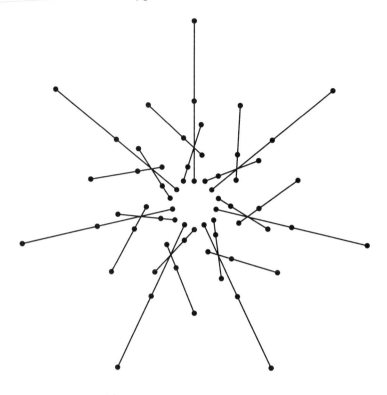

13.4   The Generalized Octagon of Order $(1, 2)$

This generalized octagon is the incidence graph of the generalized quadrangle of order $(2, 2)$ that we considered in Section 4.7. Here again is a partial list of apartments.

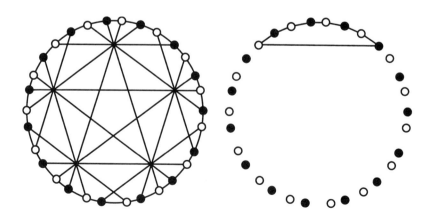

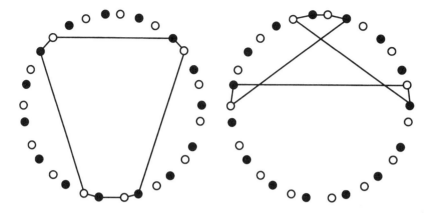

13.5   The Generalized 12-gon of Order (1, 2)

This generalized 12-gon is the incidence graph of the two generalized hexagons of order (2, 2).

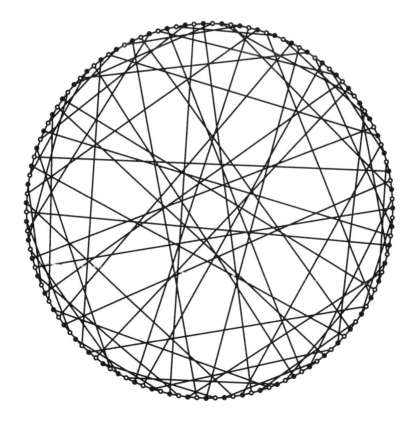

13.6 Cages

A (k,g)-cage is a regular graph of valency k, girth g, and minimum number of vertices. Some of the graphs that we have considered so far turn out to be cages.

(i) The Petersen graph is the unique $(3,5)$-cage.

If $k = q + 1$ for a prime power q, then

(ii) a $(k,6)$-cage is the incidence graph of a projective plane of order q;

(iii) a $(k,8)$-cage is the incidence graph of a generalized quadrangle of order (q,q);

(iv) a $(k,12)$-cage is the incidence graph of a generalized hexagon of order (q,q).

Furthermore,

(v) the $(k,3)$-cage is the complete graph on k vertices.

(vi) the $(k,4)$-cage is the complete bipartite graph on $2k$ vertices.

(vii) the $(2,g)$-cage is a regular g-gon.

Here is a list of $(3,g)$-cages.

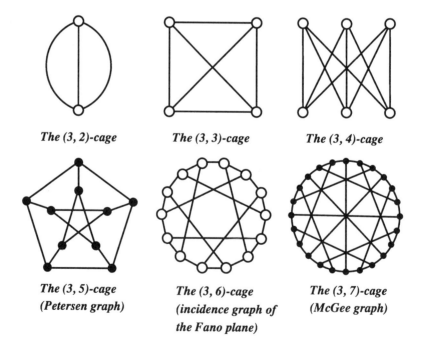

The (3, 2)-cage   The (3, 3)-cage   The (3, 4)-cage

The (3, 5)-cage   The (3, 6)-cage   The (3, 7)-cage
(Petersen graph)   (incidence graph of   (McGee graph)
                         the Fano plane)

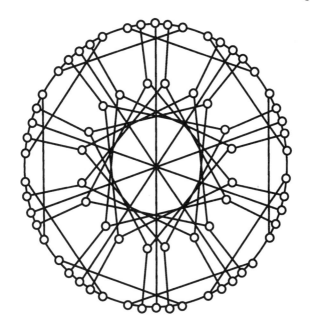

A (3, 10)-cage

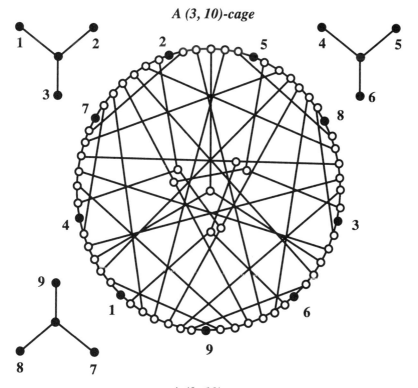

A (3, 10)-cage

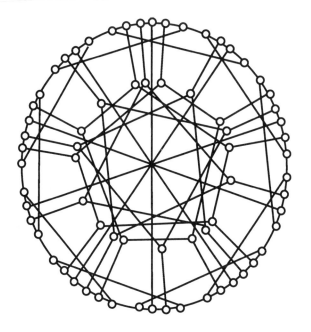

A (3, 10)-cage

For more information about cages see [14], [18, Chapter 6.9], and [53]. Here are some nice pictures of other cages. Other than the cages mentioned above and the ones depicted below, only a handful more cages are known.

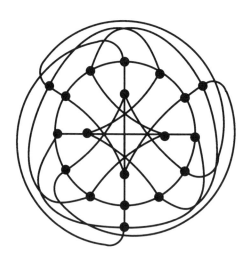

The (4, 5)-cage

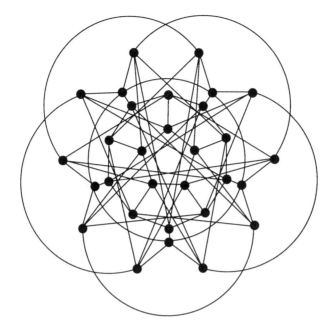

A (5, 5)-cage

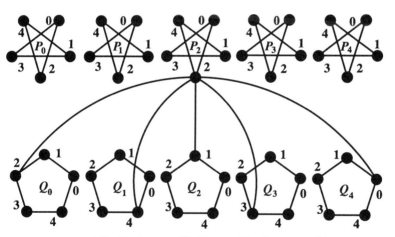

The (7, 5)-cage (Hoffmann-Singleton graph)

In the last diagram not all edges of the graph are drawn in. The graph consists of the vertices and edges of the five pentagrams $P_1, P_2, \ldots, P_5$ and the five pentagons $Q_1, Q_2, \ldots, Q_5$. Furthermore, vertex i of P_j is joined to vertex $i + jk \mod 5$ of Q_k. The resulting graph is the so-called Hoffmann–Singleton graph. This graph has some very remarkable properties. We mention only two: The union of any pentagon and any pentagram induces the

Petersen graph, and the union of three pentagons and three pentagrams induces a $(5,5)$-cage (see [18, p. 391]).

Here is another generator-only picture of the Hoffmann–Singleton graph based on a picture by K. Coolsaet. We need two generating automorphisms to generate all edges of the graph: The first one is rotating the whole picture, as usual. The second one is rotating the five small Petersen graphs simultaneously (simultaneously and through the same angle, of course).

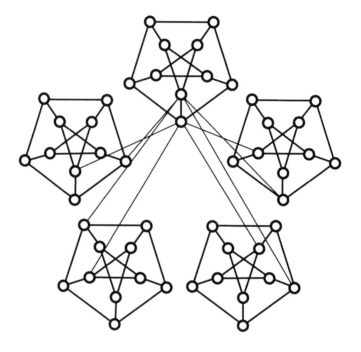

There is also a construction of this graph from the points and lines in PG$(3,2)$. We already mentioned this in Section 5.1.8.

14
Colour Pictures and Building the Models

As you can see from the color inserts that make up most of this chapter, our pictures and models look best when they are drawn in full color.

Every one of the eight colour inserts has a theme that has already been elaborated on in one of the previous chapters. We give only brief descriptions of the different inserts and refer to the respective chapters for more details.

Fano Plane

This insert shows a Singer diagram of the Fano plane and in the center of this Singer diagram the incidence graph of the Fano plane. The diagrams in the background are two copies of the traditional picture of the Fano plane. See Chapter 1 for more information about this geometry.

Configurations

This insert is all about configurations that can be modelled on a triangle such that all the rotations of the triangle translate into automorphisms of the geometries. For more information about these configurations the reader is referred to Chapter 3.

Designs

The diagram at the top of this page is a picture of the affine plane of order 3. This is followed by a Singer diagram of the Fano plane. Situated in the center of this diagram is a generator-only model of the smallest projective space. The generator-only model is also a spread of the space, that is, a partition of the space into lines. The seven images of this spread under rotation around the center of the diagram form a packing of the space, that is, a partition of the line set of the space into spreads. The red triangle in the generator-only model generates the Singer diagram of the Fano plane that surrounds the spread. The last diagram on this page is a picture of the smallest nontrivial 2-design. For more information about these geometries see Chapter 2 and Chapter 5.

Quadrangle

The large diagram in the middle of this page is the "doily," our favourite plane model of the generalized quadrangle of order $(2, 2)$, which we discussed in Chapter 4. The diagram at the bottom of the page is a "circular walk" in this geometry (see Chapter 15), and the diagram at the top of the page is the complement of this walk in the doily.

Hexagon

This picture shows one of the two generalized hexagons of order $(2, 2)$ with one of its spreads highlighted. See Chapter 13 for more information about this geometry.

Biplanes

The first two diagrams are pictures of the biplanes of orders 1 and 2. The third diagram is a generator-only model of the biplane of order 3. All three biplanes are labelled with sets of Hussain graphs. See Chapter 10.

Cages

This insert shows two of the cages described in Chapter 13. The small diagram is a picture of the famous Petersen graph, and the large diagram is a $(5, 5)$-cage. Note that the highlighted part in the middle of this cage is a "flat" dodecahedron.

Pipe Cleaner

The colour photos on the last insert show three of our three-dimensional models built from pipe cleaners: the smallest projective space, the Fano plane, and the generalized quadrangle of order $(2, 2)$. Although the diagrams in this book and these photos should give you a good idea of what these models look like, nothing beats holding one in your hand and playing

with it for a while. If you should decide to build some yourself, here are
some remarks that you may find useful.

Pipe Cleaners

Pipe cleaners are great for building models. They are flexible, yet rigid
enough to support quite a bit of weight. When you model something with
lots of straight lines, the pipe cleaner lines do not have to be absolutely
straight to look good. You can buy pipe cleaners at crafts supply shops,
sometimes news agencies (look under "party supplies"), and tobacconists.
Of course, the pipe cleaners that you buy from a tobacconist are the real
thing, but usually not as well suited to building models as the pipe cleaners
bought in crafts supply places.

Tools

You will need a small pair of pliers and a cutting tool. A pair of scissors
works as a cutting tool, but the blades will be damaged in the process of
building the model.

Time

The model of $PG(3, 2)$ is one of the most complex models. The first $PG(3, 2)$
took me about two hours to build, now, ten or so more models later, it takes
me about one hour on average.

Colours

If you want to use different colours for the different kinds of lines in the
models, be sure to match them very carefully; otherwise, you end up with
something quite awful. Have a close look at the photos, and for your first
go try to fit the colours together as I did. I find that as a general rule, it
is best to use dark colours for the lines on the outside of the models and
brighter colours for lines on the inside. This is especially true for models
of $PG(3, 2)$. Ideally, this has the effect of making the model appear to be
illuminated from inside.

Size

Size depends very much on your pipe cleaners. If you want to use a length of
pipe cleaner as a straight line, it should not bend too much when you hold
it horizontally. Just shorten the pipe cleaner until this is the case. Then
whatever is left can be used as the longest straight line in your model. The
outer edges of the model of $PG(3, 2)$ in the photo, for example, are about
15 cm long. If the model gets too small, it will start to look very crowded.
What exactly "too small" means is determined to a large extent by how
thick the pipe cleaners are that you use. If you have circles in your model,

try to make sure that you can make such a circle from one piece of pipe cleaner and hide the spot where the two ends of the pipe cleaner are spliced together under one of the "points" of the model.

Sequence of Steps in Building the Models

To build the PG(3, 2), start with the tetrahedron of reference; fill in the medians of the faces; fill in the medians of the tetrahedron and the segments connecting opposite edges of the tetrahedron; fill in the inner circles. Finally, fill in the outer circles and tie in the "points" of the model.

To build the generalized quadrangle of order (2, 2), start by building a tetrahedron; fill in the medians of the faces and the segments connecting opposite edges; remove the tetrahedron. Finally, tie in the points.

To build the Fano plane, make four circles of equal size and tie them together by three segments as in the photo; fill in the points.

Fano Plane

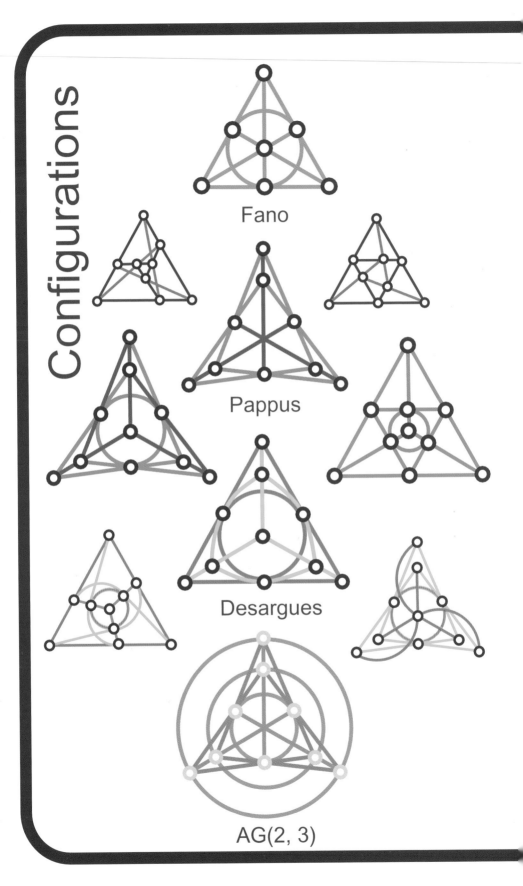

Configurations

Fano

Pappus

Desargues

AG(2, 3)

Designs

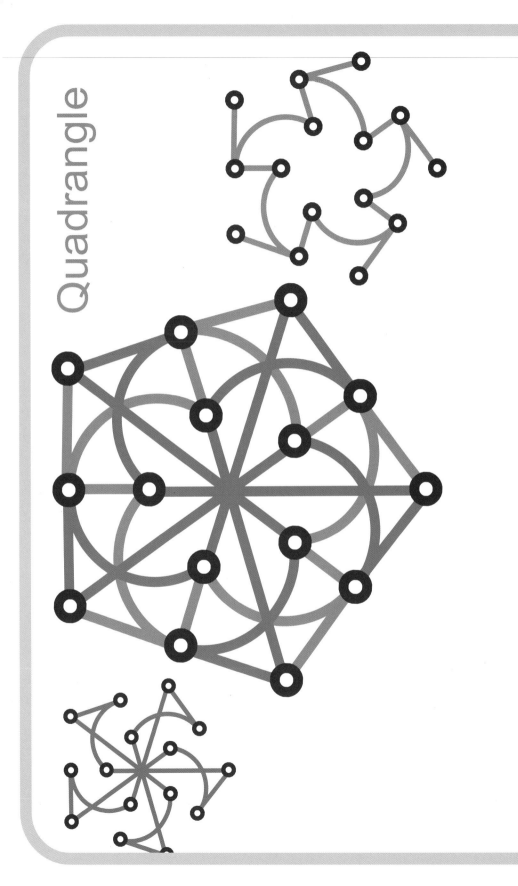

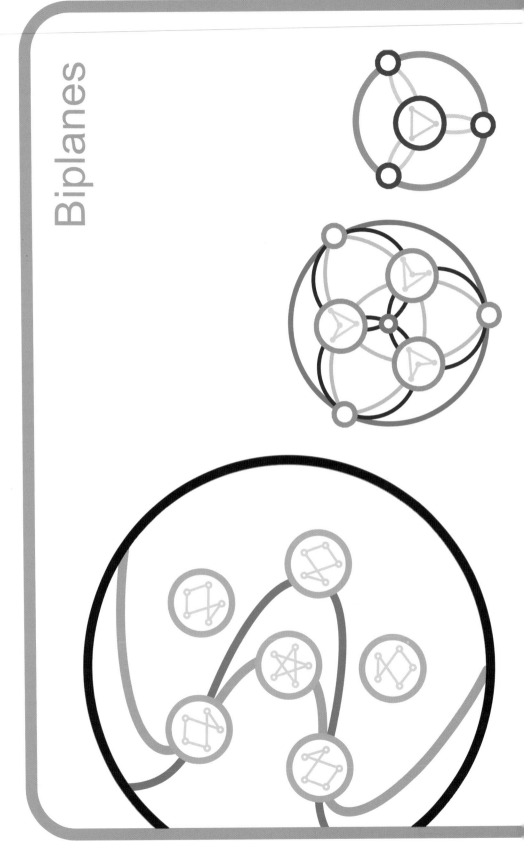

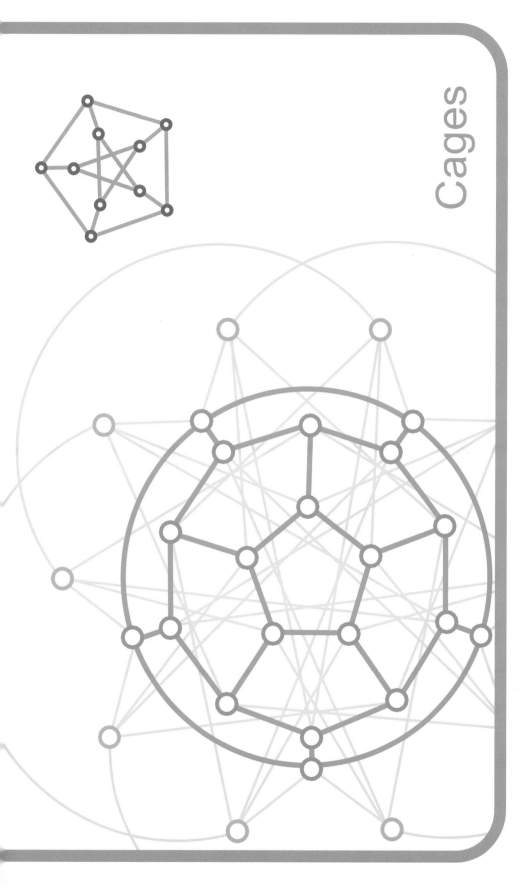

15
Some Fun Games and Puzzles

15.1 The Game "Set"—Line Spotting in 4-D

"Set—the family game of visual perception" is a card game by M. J. Falco containing 81 cards. It can be played both as a solitaire and as a multiplayer game and is highly addictive. The game has a very nice description in terms of finite geometry. Here are the rules: The 81 cards are the points of the four-dimensional affine space over the field with 3 elements. The dealer shuffles the cards/points and lays twelve of them (in a rectangle) face up on the table so that all the players/geometers can see them. The geometers try to spot sets of three points among these 12 points that are lines in our four-dimensional space. Whoever spots a line first gets to keep the three points and replaces them by three points from the stack of unused points. Since there are arcs of 12 points in our space, that is, sets of 12 points in our space that do not contain any line, it is possible to get stuck at this point. If this happens, three more points (making a total of 15) are laid up. There is always a line to be found among 15 cards (this means that maximal arcs in our space have fewer than 15 points). After removing a set of three points of a line, these are not replaced, and the number of cards is reduced to twelve again. Back to step one. Continue until all the points of the space are used up and finish the game by removing all lines from the remaining set of 12 cards. The geometer who collects the most points wins the game. It turns out that it is always possible to find a line among the last twelve points and that the game ends with either 0, 6, or 9 points left on the table.

Clearly, similar games can be played using other geometries. Still, what makes this particular setting so attractive is its practical implementation and the fact that it is very playable: Every point in our space is usually given by four coordinates. In the game these four coordinates correspond to four different aspects of a card: shape of the symbols on the card, shading of the symbols, number of the symbols, and colour of the symbols (we replace colour by background pattern). Every coordinate can be one of three things.

- Shape: oval, diamond, rectangle

- Background: no background, wave, zigzag

- Shading: solid, open, striped

- Number: one, two, three

Here is a selection of twelve cards in this game.

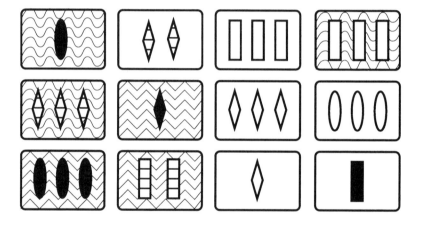

A line in our space consists of three points in which each coordinate is either the same or different for each of its points. The above set of twelve points contains a number of lines. Here are two. How many more can you find?

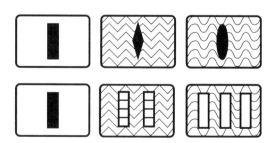

Try to find twelve points in the space that do not contain a line. Also, note that given any two points in the space, there is exactly one point that complements the two given ones into a line. Practice finding lines in the space by picking any pair of cards above and trying to figure out what the respective third card has to be.

The game can be turned into a much easier game in three-dimensional affine space over the field with three elements by restricting the game to the 27 cards with empty backgrounds, and to an even easier game in the affine plane of order 3 by restricting the game to the 9 cards with empty backgrounds that show only one symbol. Here is the point set of this affine plane. If you fill in the lines, you end up with the traditional model of the plane that we discussed before.

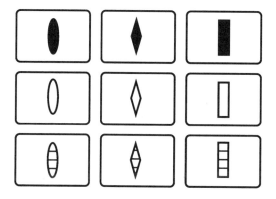

By the way, the official rules of the game do not mention its mathematical background. The company that distributes the game has a very nice web page: http://www.setgame.com/

15.2   Mill and Ticktacktoe on Geometries

In [10, p. 104] we find a description of a variation of the game mill. Instead of playing mill on the usual game board, we play it on the points of a geometry. As in the real game, the two players place pieces of two different colours on the points of the geometry. Whoever manages to have all the points of a line occupied by pieces of his own colour wins. When you fiddle around with mill on the doily, you quickly convince yourself that whoever starts the game necessarily wins. What about mill on the projective plane of order 3? What role do blocking sets play in games on geometries such as this mill variation?

Also have a go at a game or two of ticktacktoe on some of the plane configurations with parameters (p_3). Things start to get interesting with the three configurations with parameters (9_3). For two of them it is easy to show that the player who starts will always be able to win. For one of them there is a nontrivial winning strategy, again for the player who starts.

15.3   Circular Walks on Geometries

Let us go for a *circular walk* on the doily, our favourite plane picture of the generalized quadrangle of order $(2, 2)$. During our walk we will always walk along lines of the geometry. A point counts as "visited" if we used it to change directions, that is, used it to get from one line to the next. Points that we step on while walking along a line do not count as visited. Our aim is to visit every point exactly once and also use every line exactly once to move from one point to the next. Here is an example of such a walk.

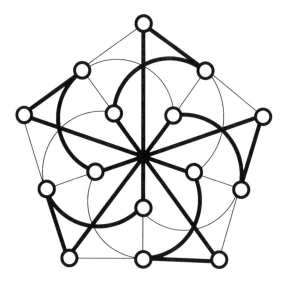

Try to complete circular walks on some of the projective planes and the two generalized hexagons. In our picture of the incidence graph of such a geometry the points and lines were arranged as vertices on a circle such that two adjacent vertices on the circle were actually joined by an edge. Therefore, the circle corresponds to the kind walk we are looking for. This, of course, means that your search will not be in vain as long as you look hard enough! On the other hand, if you succeed in finding a circular walk on a geometry, you can draw a "circular picture" of its incidence graph.

15.4   Which Generalized Quadrangles Are Magical?

The following picture on the left shows the most famous magic square.

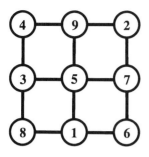 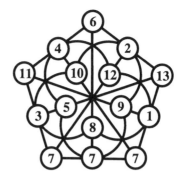

Since the 3×3 grid is also a picture of the generalized quadrangle of order $(3, 1)$, it "makes sense" to ask whether other generalized quadrangles are "magical," too. The generalized quadrangle of order $(2, 2)$ is not magical. The best we can do is to label the vertices with the consecutive numbers from 1 to 13, with the 7 right in the middle occurring 3 times, such that the sums of the labels corresponding to every single one of the lines all equal 21.

15.5 Question du Lapin

Are you familiar with the following French silhouette puzzle called the "Question du Lapin" (The "Question of the Rabbit")?

The five octagons represent five pieces of cardboard from which different shapes have been cut out: the head of a cat, the head of a horse, a vase, a flower, and a turtle. The object of this puzzle is to stack the octagons, one on top of the other, with all their vertices aligned, such that the shape of a rabbit is formed.

The pictures of the seven spreads on page 72 can be turned into a similar puzzle: Here the octagons turn into pentagons, the shapes of animals and objects turn into spreads, and the rabbit turns into a packing of the smallest projective space.

Try to assemble the following seven spreads into a packing.

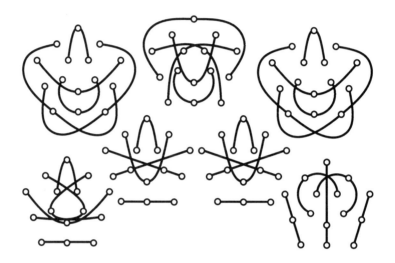

Here is another puzzle like this. This time the puzzle pieces are partial spreads of the generalized quadrangle of order $(2,2)$, and the rabbit corresponds to a packing of the doily with these partial spreads.

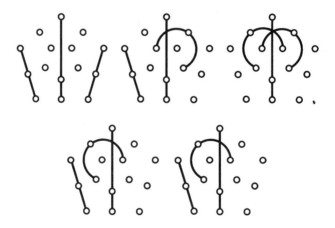

Part II

Geometries on Surfaces

16
Introduction via Flat Affine Planes

The geometries we shall be dealing with in this second part of the book are perhaps the nicest possible infinite geometries. In general, their point sets are well-known two-dimensional surfaces like the Euclidean plane, the real projective plane, the sphere, the torus, and the cylinder. Their lines, blocks, or circles are curves that are nicely embedded in these surfaces. Usually these curves will be homeomorphic to the real line or to the unit circle. The Euclidean plane, viewed as a geometry, and the geometry of circles on the unit sphere are prime examples of the kinds of geometries we will be looking at.

The standard references for geometries on surfaces are [42], [89], [90], and [103].

16.1   Some More Basic Facts and Conventions

A *topological line* on a surface homeomorphic to the xy-plane is a Jordan curve homeomorphic to an open interval that separates the plane into two open component just as a Euclidean line does. A *topological circle* on a surface is a simply closed Jordan curve on the surface. The Euclidean circles on the sphere are examples of topological circles. A *topological line* on a surface not homeomorphic to the xy-plane is a topological circle on the surface minus one of its points.

A *topological sphere* and a *topological disk* are topological spaces homeomorphic to the unit sphere and the interior of the unit circle, respectively.

16.2   The Euclidean Plane—A Flat Affine Plane

For most of us, pictures are pictures in the Euclidean plane, and letters and sentences are letters and sentences printed on a small patch of the Euclidean plane. Our pictures of finite geometries in the first part of this book are just special kinds of representations in this prototype geometry that everybody is so familiar with.

Anyway, for us the Euclidean plane is a geometry that is a so-called $\mathbf{R}^2$-*plane*, that is, a geometry whose point set is homeomorphic to the xy-plane and every one of whose lines is a topological line in the point set. Furthermore, it satisfies the Axiom (of joining) A1; that is, two points in the point set are contained in exactly one line (see Section 1.3 on affine planes).

In fact, the Euclidean plane is a *flat affine plane*, that is, an $\mathbf{R}^2$-plane that also satisfies Axiom A2 (the "parallel axiom"). Flat affine planes also automatically satisfy Axiom A3 and are therefore affine planes, as we defined them in Section 1.3.

16.2.1   Proper $\mathbf{R}^2$-Planes

It is easy to construct $\mathbf{R}^2$-planes that are not affine planes. Consider, for example, the restriction of the Euclidean plane to an open vertical strip or, more generally, to an open strictly convex region. Then we are still dealing with an $\mathbf{R}^2$-plane. Nevertheless, unless this region is the whole plane, the parallel axiom will never be satisfied. In the following two examples the two thin lines intersect and are both parallel to the thick line. This never happens in affine planes.

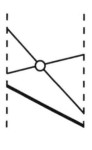
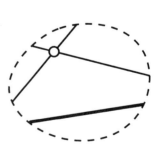

16.2.2   Pencils, Parallel Classes, Generator-Only Pictures

Here is how we usually draw a line pencil and a parallel class in the Euclidean plane.

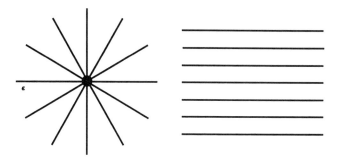

Of course, the Euclidean plane and objects like line pencils or parallel classes contain infinitely many lines. We usually draw only finitely many and rely on our imagination to fill in the rest. In fact, as in the above diagrams, we see only parts of the lines. Also, if we wanted to exhibit all the lines, we could view the parallel class as a generator-only picture for the Euclidean plane, very much as in the finite case. We generate all other lines by rotating through all possible angles. We can also regard the line pencil as a generator-only model. The rule for generating all other lines is, "Translate in all possible ways." In the finite models it was usually clear from the diagram how to generate all lines from a generator-only diagram. In our topological case things are more complicated, and a generator-only picture of a plane is usually accompanied by a rule that tells you how to generate the whole plane from the given picture.

16.2.3   The Group Dimension

In order to be able to draw good generator-only pictures of a geometry on a surface, it is necessary that the geometry have a group of automorphisms that is large in some sense and acts transitively on large chunks of the point and the line sets of the geometry. For example, in the case of the Euclidean plane we know that all translations, all rotations around a point, all dilatations, and all shears in one direction are automorphisms of the geometry. All these different kinds of automorphisms are independent of each other and generate the full group of automorphisms of the Euclidean plane. It turns out that the group of automorphisms of all the geometries we will be looking at in this part of the book are *Lie groups* of finite dimension. We call this dimension the *group dimension*. In the case of the Euclidean plane this dimension is 6. In this case the dimension is just the sum of the dimensions of the groups of translations (2), rotations (1), dilatations (2), and shears (1). Also, various subgroups of the full automorphism group act transitively on the point set, which makes it possible to draw different good generator-only pictures of our plane, as we just saw. Most of the classifications of our geometries are based on the group dimension. It turns out that a classical geometry on a surface is characterized among the geometries of the same type by the fact that it has maximal group dimension. The group dimension is a natural measure for the homogeneity of a geometry. For many types of our geometries the nonclassical geometries of maximal dimensions are known, and we will draw pictures of many of them.

16.2.4 Topological Geometries

All our geometries are *topological geometries* in that the geometric operations in the axioms that they satisfy are usually continuous. For example, in a flat affine plane the connecting line of two points changes its shape and position continuously as the two points are moved about continuously.

16.2.5 The Way Lines Intersect

If two lines in an $\mathbf{R}^2$-plane intersect, they *intersect transversally*, just like lines in the Euclidean plane; they never just *touch* as in the picture on the right.

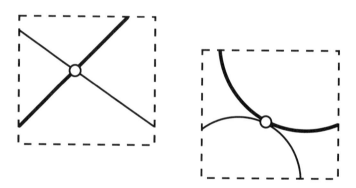

Remember also that Axiom A1 implies that two lines never intersect in more than one point. Most of the other kinds of geometries on surfaces we will be dealing with also satisfy some kind of axiom of joining that guarantees that n points in "general position" are contained in a unique line/circle. In such a plane two lines/circles intersect in at most $n - 1$ points, and if they intersect in this maximal number of points, then they intersect transversally in every single one of the points. For example, for the geometry of circles on the sphere, n equals 3 and "general position" means that the 3 points that are to be connected are distinct. So, if two circles intersect in 2 points, they intersect as shown in the diagram on the left. If two circles intersect in fewer than $n - 1$ points, then they can touch in some of these points as in the diagram on the right.

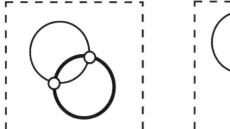 

16.2.6   Ovals and Maximal Arcs

An *arc* in an affine plane or a projective plane is a set of points such that no three points in the set are collinear. An arc is called *maximal* if it is not contained in a bigger arc. Examples for maximal arcs are ovals in finite projective planes of odd order and hyperovals in finite projective planes of even order (see Section 1.6 for a definition of ovals in affine and projective planes). The Euclidean circles and ellipses are the nondegenerate conic sections in the Euclidean plane and as such are examples of *topological ovals*, that is, ovals that are topological circles. It is possible to construct ovals and hyperovals in the Euclidean plane that are not topological by a process called *transfinite induction*. The points in the ovals and hyperovals constructed in this manner are scattered in a wild manner all over the plane and do not play a significant part in the theory of flat affine planes.

The topological ovals in the Euclidean plane are just the differentiable, strictly convex, simply closed curves in the plane. Topological ovals can be shown to exist in all flat affine planes (see [88]). In many ways they behave exactly like ovals in the Euclidean plane. For example, there always exist exterior lines, a secant line intersects a topological oval transversally in its two points of intersection, and a tangent line really touches the oval in its point of intersection as shown in the following picture on the right.

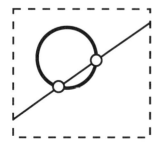 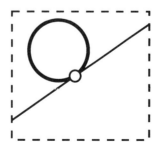

Given a point not on a topological oval, it is always possible to find a secant of the oval that contains the point. This means that a topological oval is always a maximal arc. See [20] and [90] for information about topological ovals.

Nice Maximal Arcs That Are Not Topological Ovals

Here are three examples of maximal arcs made up of a number of convex curves each.

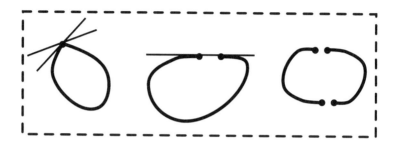

Note that the first set is homeomorphic to the circle, and no three of its points are collinear. Still, it fails to be an oval, since there are many lines that contain only the distinguished corner. This means that it does not satisfy Axiom O2 (see Section 1.6). The second set is no longer a topological circle. Actually, it is supposed to be homeomorphic to a closed interval, and the connecting line of its two end points touches the curve in both points in the analytical sense. It also does not satisfy Axiom O2.

16.3 Nonclassical $\mathbf{R}^2$-Planes

In this section we present some of the most appealing constructions for nonclassical flat affine planes and $\mathbf{R}^2$-planes.

16.3.1 Moulton Planes

The Moulton planes were introduced by Moulton [71] at the beginning of the twentieth century and are some of the earliest examples of nonclassical flat affine planes. Like the Euclidean plane, they contain the vertical lines. The remaining lines can be generated from one line pencil. Here is what this line pencil looks like. It contains the vertical and all the straight lines through the origin with nonpositive slope. Furthermore, it contains all "bent" lines through the origin that start out as straight lines of positive slope a to the left of the origin and continue as straight lines of slope ka to the right of the origin. Here k is a fixed positive number. The generating rule is, "Translate in the vertical direction."

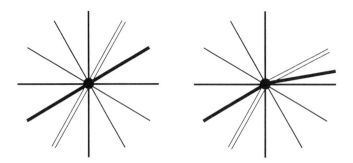

You can also think of this plane as being glued together along the y-axis from two Euclidean halves. The number k is the "glue" factor. The resulting plane is the Euclidean plane only if $k = 1$. Two Moulton planes with glue factors k and l are isomorphic only if $k = l$ or $k = 1/l$. The verticals form one parallel class. The remaining parallel classes correspond to the nonvertical lines in the generating line pencil; take any line in the pencil and all its vertical translates and you get a parallel class in the Moulton plane. There are three essentially different line pencils depending on whether the carrier of the pencil is situated to the right, on, or to the left of the y-axis.

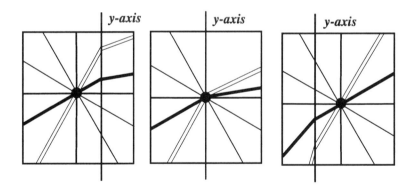

See also [97] for a vast generalization of Moulton's construction.

16.3.2 Shift Planes

We arrive at one of the so-called parabola models of the Euclidean plane by deforming it and all its lines by the following homeomorphism:

$$\mathbf{R}^2 \to \mathbf{R}^2 : (x, y) \mapsto (x, y + ax^2).$$

Here a is a fixed nonzero number. In the resulting representation the lines of the plane are the verticals plus all translates of the *Euclidean parabola* that corresponds to the function $x \mapsto ax^2$. Here is a generator-only picture for this plane. The generating rule is, "Translate in all possible ways."

And here are the different parallel classes.

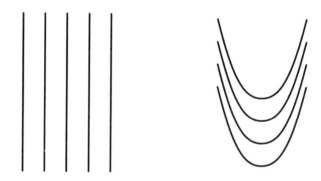

Note that there is only one parallel class consisting of straight lines and infinitely many different ones consisting of parabolas.

Nonclassical flat affine planes can be constructed by replacing the parabola in this construction by a curve that looks like a parabola. Any differentiable curve whose derivative is a homeomorphism of $\mathbf{R}$ to itself will do. An example is the graph of the function $x \mapsto x^4$. But the curves do not have to be differentiable. The only requirements are that the curve be strictly convex (or concave) and that it be "steeper" than any linear function. The resulting plane will be just another model of the Euclidean plane if and only if the curve is a Euclidean parabola. Planes constructed in this manner are called *shift planes*. Note that in a shift plane, just as in the Euclidean plane, all line pencils look the same, that is, are translates of each other. Here is what a pencil looks like.

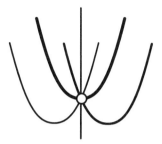

We call topological lines that generate shift planes *topological parabolas*. It turns out that these curves are exactly those that can be completed by the infinite point of the verticals, resulting in topological circles in the projective completion of the Euclidean plane that are maximal arcs in this projective plane.

See [90] and [39] for more information about this construction.

16.3.3   Arc Planes

Arc planes are $\mathbf{R}^2$-planes that are generalizations of shift planes. A strictly convex topological line l is an arc in the Euclidean plane. We call such a line a *topological arc*. We call a topological arc *unbounded* if it is not contained in a strip bounded by two parallel Euclidean lines. Let A be a finite collection of unbounded topological arcs such that given a secant line of one of the arcs in the set, none of the other arcs in the set has any parallels of this line as a secant line. Let S be the collection of all Euclidean lines not parallel to a secant line of one of the arcs in A. Then the collection of all translates of arcs in A together with all Euclidean lines in S is an $\mathbf{R}^2$-plane. Here are some examples for A and S.

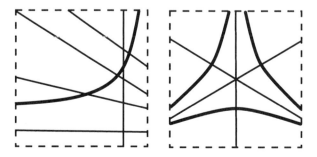

In the first example, A is supposed to contain the graph of the exponential function, and S contains all Euclidean lines with nonpositive slope. In the second example, A contains three branches of Euclidean hyperbolas, any two of which share a common asymptote. Here S consists of all parallels to the three asymptotes.

See [38] and [41] for more details about this construction.

16.3.4 Integrated Foliations

Note that the nonvertical lines in the Euclidean plane are the graphs of all linear functions $x \mapsto ax + b$, $a, b \in \mathbf{R}$. The linear functions are the integral functions of the constant functions $x \mapsto a$, $a \in \mathbf{R}$. Each of these constants in turn corresponds to one of the parallel classes of the Euclidean plane; that is, a set of curves that partitions the point set. In this way the Euclidean plane "is the integral" of the parallel class consisting of the horizontal lines. If we replace the parallel class in this construction by the parallel class consisting of all lines with slope $k \neq 0$, then the "integral" of this parallel class is one of the parabola models of the Euclidean plane that we considered above. More generally, given a set of functions whose graphs partition the plane, like this,

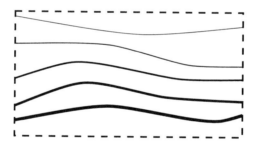

we can define a geometry whose point set is the whole plane and whose lines are the verticals plus the graphs of all integral functions of the functions in the set. This geometry is always an $\mathbf{R}^2$-plane, and if any two of the elements in one of the partitions stay "far enough apart," this geometry is a flat affine plane.

See [75] for details about this construction.

16.3.5 Gluing Constructions

Gluing Between Two Parallel Lines

Consider two flat affine planes that live on the same point set and that share a pair of parallel lines. For example, the Euclidean plane and one of the shift planes have all the verticals in common. Choose two such parallel lines as in the following diagram.

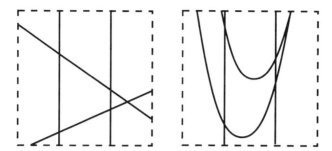

Then we can replace everything in the strip between the two lines on the right by everything in the corresponding strip on the left. Two half-lines on the left and right of the strip that come from the same line get extended by the unique (by Axiom A1) line segment inside the strip that connects the two end points of the half-lines. The resulting geometry is also a flat affine plane, and it looks as follows.

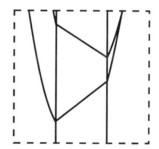

Gluing with Respect to Topological Ovals

It is also possible to glue flat affine planes together, in very much the same way as above, if they share a topological oval.

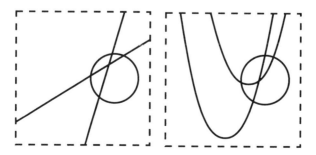

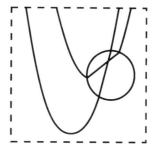

See [40], [86], and [111] for more information about these gluing constructions.

All Flat Affine Planes are Deformed Euclidean Planes

It is possible to use a gluing construction to continuously deform any given flat affine plane A into the Euclidean plane as follows: Use a homeomorphism from the point set of A to the point set of the Euclidean plane to make A into a flat affine plane that shares the point set and all vertical lines with the Euclidean plane. Now we can produce new planes by gluing with respect to any two of the verticals (here the plane A takes the place of the shift plane in the above example). Let $-v(t)$ and $v(t)$, $t \in \mathbf{R}^+$, be the two verticals at distance t from the y-axis, and A_t the flat affine plane associated with these two verticals. As we let t go from 0 to infinity, the plane A, which is the plane A_0, is continuously deformed into the Euclidean plane. The two verticals moving further and further apart as t gets larger remind us of a theater curtain opening. At first we see A. Then the curtain opens to reveal ever larger chunks of the Euclidean plane until finally, when the curtain is open all the way, only the Euclidean plane is visible, and someone takes a bow. See [81] for more details.

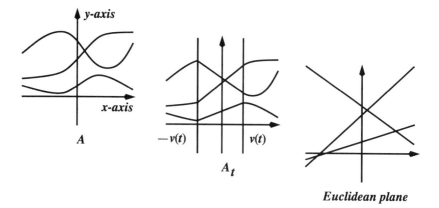

Euclidean plane

16.4   Classification

All $\mathbf{R}^2$-planes with group dimension at least 3 are known. The Euclidean plane is the only plane with group dimension at least 5; it has group dimension 6. The planes with group dimension 4 are the restriction of the Euclidean plane to the upper halfplane and the flat affine planes we arrive at by removing the extended y-axis from the projective extensions of the nonclassical Moulton planes. There are two classes of flat affine planes with group dimension 3. First, the so-called *Cartesian planes* which include the nonclassical Moulton planes. Second, the *skew parabola planes*, that is, special shift planes generated by the functions

$$\mathbf{R} \to \mathbf{R} : x \mapsto \begin{cases} x^d & \text{for} \quad x \geq 0 \\ c(-x)^d & \text{for} \quad x \leq 0 \end{cases}$$

where $0 < c \leq 1 < d$, $(c,d) \neq (1,2)$. Proper $\mathbf{R}^2$-planes with group dimension 3 also exist. First, there is the hyperbolic plane, that is, the restriction of the Euclidean plane to the interior of the unit circle.

Then there are four arc planes generated by the graphs of the following functions:

- x^s for $x > 0$, where $s \leq -1$;

- x^s and rx^s for $x > 0$, where $r \leq -1$ and $s < 0$;

- e^x for $x \in \mathbf{R}$;

- e^x and $-\text{sgn}(s)e^{sx}$ for $x \in \mathbf{R}$, where $s \leq -1$ or $s > 1$.

Furthermore, there is the so-called *Strambach* SL_2-*plane* and two families of planes whose groups fix precisely one line.

The flat affine planes with group dimension at least 2 have also been classified. It is hopeless to try to classify the $\mathbf{R}^2$-planes with group dimension 1. It is possible to construct *rigid* $\mathbf{R}^2$-planes, that is, planes that do not admit any automorphisms. The restriction of the Euclidean plane to the interior of a sufficiently nonsymmetric topological oval is such a rigid plane. Rigid flat affine planes are easily constructed using the gluing constructions. For details about this classification see [42] and [90].

16.5 Semibiplanes

In Chapter 11 we constructed semibiplanes from projective planes that admit involutory automorphisms. This construction can be translated into a construction of semibiplanes from affine planes. Let i be an involutory automorphism of an affine plane with exactly one fixed point. Then a semibiplane can be constructed as follows: Let the points be the unordered pairs of distinct points of the affine plane that get exchanged by i. Similarly, let the lines be the unordered pairs of distinct lines of the affine plane that get exchanged by i. The resulting semibiplane is divisible. The parallel classes of lines in the semibiplane correspond to the parallel classes in the affine plane, and the parallel classes of points correspond to the lines through the fixed point.

Take for example the reflection of the xy-plane through the origin. This reflection is an automorphism of the Euclidean plane. It fixes only the origin and the lines through the origin. So, points of the associated semibiplane are pairs of points at equal distance from the origin whose connecting line contains the origin. The lines of the semibiplane are pairs of parallel lines of equal distance from the origin. Here is an example of two nonparallel points and their associated two nonparallel connecting lines.

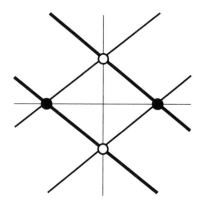

17
Flat Circle Planes—An Overview

In this section we give an overview of the most important geometries on surfaces whose lines/circles are topological circles.

17.1 The "Axiom of Joining" and the "Map"

In the "map of circle planes" on the next page, the different types of geometries are represented by one icon each. Actually, you have to think of this map as being continued to the right as indicated. So, we are really talking about an infinite number of different geometries.

As you can see, the surfaces these geometries are living on are the Möbius strip, the cylinder, the sphere, the torus, and the real projective plane. We call two points on one of these surfaces *parallel* if:

- they are identical points on the real projective plane or the sphere;

- they are contained in the same vertical line on the cylinder or the Möbius strip;

- they are contained in the same vertical or the same horizontal circle on the torus.

Parallel points are never contained in a circle. If n is the number of dots you see in the middle of an icon, the geometry corresponding to the icon satisfies the following *axiom of joining:*

Axiom of joining for flat circle planes

(J) Every n pairwise nonparallel points are contained in a unique circle.

We call n the *rank* of the geometry.

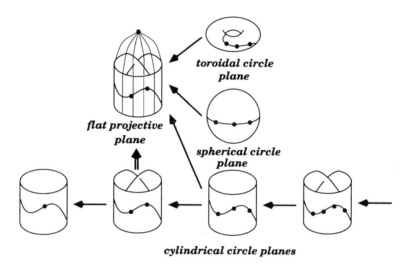

cylindrical circle planes

The different types of geometries are grouped around the following *classical examples*:

- *Flat projective planes:* The classical projective plane associated with the field of real numbers.

- *Spherical circle planes:* The geometry of nontrivial plane sections of the unit sphere, that is, the geometry of Euclidean circles on this sphere.

- *Toroidal circle planes:* The geometry of nontrivial plane sections of a hyperbolic quadric in real projective three-space.

- *Cylindrical circle planes of rank n:* The (topological) projective closure of the geometry that corresponds to the set of polynomials of degree at most $n - 1$ over the real numbers. These circle planes have interpretations as periodic and antiperiodic n-unisolvent sets of functions (see [78]). In particular, the classical cylindrical circle plane of rank 3 can also be represented as the geometry of nontrivial plane sections of a cylinder over a circle in the xy-plane. Similarly, the classical

cylindrical circle plane of rank 1 can be considered as the geometry
of horizontal plane sections of a cylinder over a circle in the xy-plane.

Flat projective planes are really projective planes as we defined them in
Section 1.2. In particular, Axiom J, which corresponds to Axiom P1 for
projective planes, implies the other two axioms P2 and P3.

It is the three different types of flat circle planes of rank 3 that are usually
referred to as *the* flat circle planes. Their classical examples arise as the
geometries of nontrivial plane sections of the three different kinds of nondegenerate quadrics in real projective three-space and are representatives of
the three different kinds of Benz planes as we defined them in Chapter 12.

17.2   Nested Flat Circle Planes

It is possible that a flat circle plane has a much richer local structure than
implied by Axiom J. To every point of one of the classical circle planes
of rank $n > 2$, for example, there is associated a classical circle plane of
rank $n - 1$. The single arrows in the above map indicate what type of
circle plane this associated plane is. Consider, as a concrete example, the
geometry of circles on the sphere. This is a flat circle plane of rank 3. The
derived geometry at one of its points is the Euclidean plane, and the flat
circle plane associated with the point is the projective extension of the
Euclidean plane, that is, the classical flat projective plane. All this means
that the classical circle planes have a nested structure. If a flat circle plane
has a local structure similar to its classical counterpart, we call it *nested*.
All flat circle planes of rank at most 2 turn out to be nested. Not all flat
circle planes of higher rank are nested, and the ones of rank 3 that are
have special names: Nested toroidal circle planes are called *flat Minkowski
planes*, nested spherical circle planes are called *flat Möbius planes*, and
nested cylindrical circle planes of rank 3 are called *flat Laguerre planes*.
These geometries are really Benz planes as we defined them in Chapter 12.

Here is a precise definition of what we mean by a nested flat circle plane.
Let C be a flat circle plane and let p be one of its points.

The *first restricted geometry* C_p of C at p has as its point set the set of
all points not parallel to p. Its lines are the circles containing p that have
been punctured at p.

The *second restricted geometry* C^p of C at p has as its point set the set
of all points not parallel to p. Its lines are the circles of C that have been
punctured at points parallel to p.

Let V_p be the set of all parallel classes not containing p that have been
punctured at points parallel to p. Let $\bar{C}_p$ and $\bar{C}^p$ coincide with C_p and C^p,
respectively, except that the elements of V_p are also considered as lines. The
geometry $\bar{C}_p$ is also called the *complete derived geometry at the point p*.

Let C be a cylindrical circle plane of rank n. A cylindrical circle plane D of rank $n-1$ is called a *derived plane* of C at p if there exists a point $q \in D$ such that C_p is *topologically isomorphic* to D^q, that is, there is a homeomorphism of the point set of C_p to the point set of D^q that induces an automorphism of the two geometries. If a derived cylindrical plane exists, then it is uniquely determined up to topological isomorphism (see [78]). It therefore makes sense to speak of the derived cylindrical circle plane at a point.

The cylindrical circle plane C is *nested* if either $n = 1$ or if the derived cylindrical circle planes at all its points exist and the derived planes are nested themselves.

Let $n = 2$. Then it is easy to prove that C is automatically nested and that the geometry $\bar{C}^p$ is a flat affine plane, that is, a flat projective plane minus a line and its points. Also, C itself is just a flat projective plane minus a point and the lines through it. This projective plane can be reconstructed by one-point compactifying the point set of C by a special point (this is indicated by the double arrow in the map). The lines of this projective plane are the circles of C and the parallel classes of points in C to every single one of which the special point has been adjoined. Conversely, by removing any point p from a flat projective plane, we arrive at a cylindrical circle plane of rank 2 in the obvious way.

We want to define what it means for the other types of flat circle planes to be "nested." From the above remark it makes sense to say that every flat projective plane is nested, that is, every flat circle plane of rank less than 3 is nested. A flat circle planes of rank 3 is nested if the complete derived geometries at all its points are flat affine planes, that is, essentially flat projective planes. Note that for a cylindrical circle plane of rank 3 both definitions of nested apply and describe the same class of geometries.

17.3   Interpolation

Clearly, Axiom J has a lot to do with Lagrange interpolation. In fact, the circle set of a cylindrical circle plane of rank n can be interpreted as a system of continuous periodic or antiperiodic functions that solves the Lagrange interpolation problem of order n, depending on whether n is odd or even. Also, for a cylindrical circle plane to be nested just means that it solves a topological version of the Hermite interpolation problem. In particular, a system of $(n - 1)$-times continuously differentiable periodic or antiperiodic functions that solves the Hermite interpolation problem of order n corresponds to a nested cylindrical circle plane of rank n (see [78] for more details).

18
Flat Projective Planes

In this chapter we concentrate on flat projective planes. Just to reiterate, a *flat projective plane* is a geometry whose point set is the real projective plane (considered as a topological space only), whose lines are topological circles, and that satisfies the Axiom (of joining) P1 for projective planes (see Section 1.2). Every such geometry automatically also satisfies the other two axioms for projective planes; that is, it is automatically a projective plane. Also, there are essentially two different ways in which a topological circle can be embedded in the real projective plane. In one kind of embedding the curve is embedded such that it separates the surface into two open components; one is homeomorphic to the unit disk, the other one to the Möbius strip. In the other kind of embedding the curve does not separate the surface. All lines in a flat projective plane are embedded like this.

By removing a line from a flat projective plane, we arrive at a flat affine plane. On the other hand, it can be shown that every flat affine plane arises from a uniquely determined flat projective plane in this manner. Of course, as a geometry this flat projective plane is the same as the projective completion of the flat affine plane. This implies that the projective completions of our examples of nonclassical flat affine planes are nonclassical flat projective planes.

As a surface, the real projective plane cannot be embedded in Euclidean space without self-intersection. Still, there are some nice representations of this surface and the flat projective planes living on it. The following diagram shows some of the nicest such representations. We first give a detailed description of how these representations come about in the case of the projective plane over the real numbers. Every single one of the representations gives some special insight into the nature of this plane and gives rise to ways in which to construct nonclassical projective planes.

Following this we summarize the classification of the flat projective planes with group dimension greater than 1.

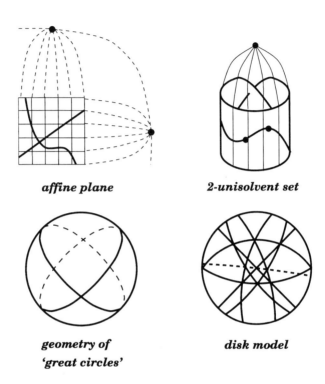

affine plane   *2-unisolvent set*

geometry of   *disk model*
'great circles'

18.1   Models of the Real Projective Plane

We have already mentioned one popular representation of the classical projective plane over a field in Section 1.2. In the case of the real numbers $\mathbf{R}$, the points and the lines of the geometry are the lines and the planes through the origin of $\mathbf{R}^3$, respectively.

Associated with every surface S in Euclidean space there are various kinds of geometries associated with this model. The lines of these geometries are the intersections of the planes containing the origin with the surface. Its points are either the points of the surface, the sets of intersection of the lines through the origin with the surface, or a mixture of these two different kinds of points. The most popular representations of the real projective plane are constructed in this manner. Let us go through a list of surfaces, the associated models, and their meaning.

18.1.1   The Euclidean Plane Plus Its Line at Infinity

A plane different from but parallel to the xy-plane: The lines of the associated geometry are the Euclidean lines contained in the plane; its points are points of the plane. Well, this is just the Euclidean plane. Its line at infinity corresponds to the xy-plane and the points at infinity to the lines through the origin in this plane.

18.1.2   The Geometry of Great Circles

The unit sphere: There are two geometries associated with this surface. Both have as their lines the great circles on the sphere. The points of the first geometry are pairs of antipodal points on the sphere. This geometry is another popular representation of the real projective plane.

The points of the second geometry are the points of the sphere. The resulting geometry is called the *geometry of great circles* or the *double cover of the real projective plane*—"double cover" because every point of the projective plane corresponds to two points in this geometry. Also, as a surface, the sphere is a topological double cover of the real projective plane with the antipodal map as the identifying map.

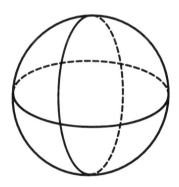

In general, a geometry of topological circles on the sphere is a *double cover of a flat projective plane* if there exists a fixed-point-free involutory orientation-reversing homeomorphism i of the sphere to itself such that (1) i is an automorphism of the geometry that globally fixes every topological circle in the geometry; (2) two points that are not exchanged by i are contained in a unique circle in the geometry. If we topologically identify the sphere via i, we arrive at a surface homeomorphic to the real projective plane. The set of topological circles on the sphere turns into a set of topological circles on this surface that is the line set of a flat projective plane. In this way a double cover is a special representation of this particular flat projective plane. See [87] for more information about double covers.

18.1.3 Möbius Strip—Antiperiodic 2-Unisolvent Set

A cylinder over the unit circle: The lines of this geometry are pairs of verticals plus a selection of ellipses. Let the points of the geometry be the points of the cylinder. This geometry is essentially the same as the geometry of great circles. The only difference is that we are missing one pair of antipodal points.

If we parametrize the unit circle by $[0, 2\pi)$ in the usual way, the non-vertical lines are the graphs of the periodic functions of the form

$$[0, 2\pi] \to \mathbf{R} : t \mapsto a \sin t + b \cos t, a, b \in \mathbf{R}.$$

We identify antipodal points of the cylinder. This gives a representation of the classical flat projective plane on the Möbius strip M, which we consider to be the strip $[0, \pi] \times \mathbf{R}$, whose left and right boundaries have been identified via the map

$$\{0\} \times \mathbf{R} \to \{\pi\} \times \mathbf{R} : (0, y) \mapsto (\pi, -y).$$

In this representation the point set consists of the points in M. The lines are the verticals and the graphs of the functions

$$[0, \pi] \to \mathbf{R} : t \mapsto a \sin t + b \cos t, a, b \in \mathbf{R}.$$

Note that the graph of a function representing a nonvertical line is a topological circle on the Möbius strip. The set of functions representing the nonvertical lines is an *antiperiodic 2-unisolvent set of continuous functions*. Here "antiperiodic" means that $f(0) = -f(\pi)$ for all functions f in the set, and 2-unisolvent means that given two points (x_1, y_1) and (x_2, y_2) on the Möbius strip such that $x_1 \neq x_2$, there is a unique function in the set that interpolates both points.

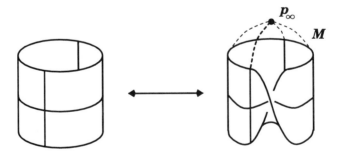

In general, given an antiperiodic 2-unisolvent set of continuous functions, consider the geometry on the Möbius strip associated with it consisting of all verticals and the graphs of the functions in the set. The geometry we arrive at by one-point compactifying the Möbius strip by a point p_∞

and extending all verticals by this point is a flat projective plane. In this way an antiperiodic 2-unisolvent set of continuous functions is a special representation of this particular flat projective plane.

18.1.4 A Disk Model

The upper (unit) hemisphere plus its equator (the unit circle): Topologically speaking, this surface is a disk. The points of the associated geometry are the points of the upper hemisphere plus pairs of antipodal points on the unit circle. This is a representation of the real projective plane that is "half" of the geometry of great circles. We project the surface from above onto the unit disk in the xy-plane to arrive at a representation of the real projective plane on the closed unit disk. Its points are the interior points of the disk plus all pairs of antipodal points on the unit circle. Its lines are halves of ellipses touching the unit circle in antipodal points plus the line segments connecting antipodal points plus the unit circle itself. The diagram on the right is a generator-only model of this plane. The generation rule is, "Rotate around the origin in all possible ways." This diagram is very reminiscent of the picture of the projective plane of order 3 on page 99. The restriction of this geometry to the interior of the disk is a representation of the Euclidean plane. The line at infinity of this Euclidean plane is the unit circle. What are its parallel classes?

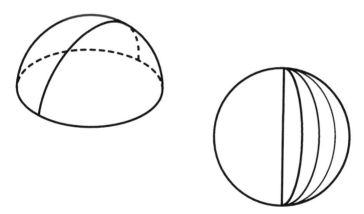

In general, a *disk model of a flat projective plane* is a geometry on a *topological disk* whose boundary c is provided with a fixed-point-free involutory orientation-reversing homeomorphism i. Its points are the interior points of the disk together with the set of pairs of boundary points that get exchanged by i. Its line set consists of c and a collection of Jordan curves whose two end points get exchanged by i. Furthermore, the geometry satisfies the Axiom (of joining) P1. The topological space we arrive at by identifying the boundary of the topological disk via i is a surface homeomorphic to the real projective plane. The line set of our geometry turns into a set of topological circles on this surface that is the line set of a flat projective plane. In this way a disk model is a special representation of this particular flat projective plane. See [77] for more information about disk models.

18.2   Recycled Nonclassical Projective Planes

Above we gave a disk model of the real projective plane on the unit disk. Here is another such model, again constructed on the unit disk. Let the points of the geometry coincide with the points of the classical disk model, that is, the interior points plus all pairs of points on the unit circle that get exchanged by the antipodal map. Let the lines be the Euclidean line and circle segments connecting antipodal points on the unit circle, plus the unit circle. The following diagram on the left is a generator-only model, and the diagram on the right shows a Pappus configuration in this model of the classical flat projective plane.

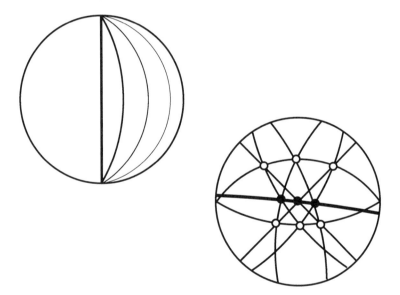

If we replace the antipodal map in this construction by an arbitrary fixed-point-free orientation-reversing involutory homeomorphism of the circle, then the constructed geometry is always going to be a disk model of a flat projective plane. In general, projective planes constructed like this will be nonclassical. Consider the above diagram on the right. It is easy to construct a suitable involutory homeomorphism that, just like the antipodal map, exchanges the ends of the thin segments but does not exchange the end points of the thick segment. This means that the associated projective plane still contains the thin part of the configuration. It does not contain the thick segment. Since the three solid points are contained in exactly one Euclidean circle, the Pappus configuration does not close in this plane. Hence the flat projective plane under discussion is not classical.

The *bundle involution* of the unit circle associated with a point p not on the circle is defined as follows.

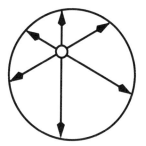 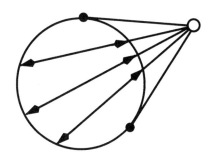

In particular, the bundle involution associated with an interior point of the unit circle exchanges the two points of intersection of a line through this point with the circle. All bundle involutions associated with interior points are "suitable" for our construction and are exactly the ones that yield flat projective planes isomorphic to the classical projective plane. Note that the antipodal map is the bundle involution associated with the origin of the plane.

See [77] for more information about this construction.

18.3   Salzmann's Classification

Salzmann classified the flat projective planes with group dimension at least 3. It turns out that the only flat projective plane with group dimension at least 5 is the classical projective plane; it has group dimension 8. The flat projective planes with group dimension 4 are the projective extensions of the nonclassical Moulton planes. There are three classes of flat projective planes with group dimension 3. These are the so-called skew hyperbolic planes, the projective extensions of the skew parabola planes, and the projective extensions of Cartesian planes. The planes with group dimension at least 2 have also been classified. It is hopeless to try to classify flat projective planes with group dimension 1. It is possible to construct *rigid* flat projective planes, that is, planes that do not admit any automorphisms. Many of the recycled projective planes above are rigid planes.

For details about this classification see [42] and [90].

19
Spherical Circle Planes

19.1   Intro via Ovoidal Spherical Circle Planes

In this chapter we concentrate on spherical circle planes. Just to reiterate, a *spherical circle plane* is a geometry whose point set is homeomorphic to the sphere and all of whose circles are subsets of the sphere that are topological circles. The axiom of joining that every spherical circle plane satisfies guarantees a unique circle that contains three given points. The derived geometry of a spherical circle plane at a point is an $\mathbf{R}^2$-plane (check this!). A spherical circle plane is called a *flat Möbius plane* or *flat inversive plane* if the derived geometries at all its points are flat affine planes. It is easy to show that a flat spherical circle plane is a flat Möbius plane if and only if it satisfies the so-called axiom of touching: Given a point p on a circle c and another point q, there is a uniquely determined circle d that contains both points and touches c, that is, intersects c only in p or coincides with c.

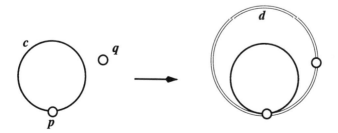

In the case of a rank 2 geometry like an affine or projective plane, we usually verify the axiom of joining by checking that all possible line pencils partition the plane (minus the respective carriers of the pencils). Similarly, when we want to check the axiom of joining in a rank 3 circle plane, we look at all possible pencils of circles through two points, and we check the axiom of touching by looking at all possible *tangent pencils*. Again, we have to check that all these circle pencils partition the plane.

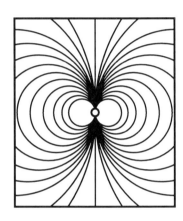

The classical example of a flat Möbius plane is the geometry of nontrivial plane sections of the unit sphere, or equivalently, the geometry of Euclidean circles on the unit sphere. In general, given a strictly convex topological sphere in Euclidean space, the geometry of nontrivial plane sections of this surface is a spherical circle plane. These kinds of circle planes are also called *ovoidal spherical circle planes*. An ovoidal circle plane is a flat Möbius plane if and only if the topological sphere is differentiable. If it has a "corner," the derived geometry at this corner is no longer an affine plane, and the spherical circle plane is not a flat Möbius plane.

Consider a spherical circle plane on the unit sphere and the tangent plane of the sphere at its south pole. We stereographically project the sphere from its north pole onto this plane. This map is a homeomorphism of the sphere minus its north pole onto the plane. Under this stereographic projection, circles that do not contain the north pole are mapped onto topological circles of the plane, and circles that contain this point onto topological lines. The geometry on the plane whose lines are the curves of the last kind form a model of the derived geometry of our spherical circle plane at its north pole. If the spherical circle plane under discussion is the classical example, then circles of the first kind are mapped onto Euclidean circles in the plane, and circles of the second kind are mapped onto Euclidean lines.

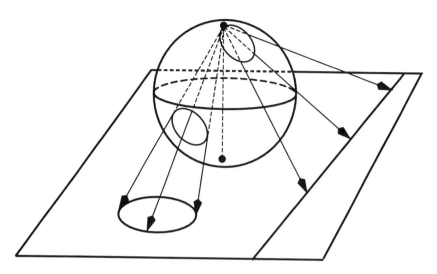

This shows that the derived geometry of this classical example at any of its points is the Euclidean plane. Convince yourself, using similar stereographic projections, that the derived geometry at any point of an ovoidal spherical circle plane is a geometry that is a restriction of the Euclidean plane to one of its open, convex subsets and that this subset is the full plane if the circle plane is a flat Möbius plane.

For more detailed information about spherical circle planes see [103], [105], [106], and [120]. For other important aspects of the classical flat Möbius planes see [4] and [122]. See Chapter 12 for information about finite Möbius planes.

19.2   The Miquel and Bundle Configurations

The Miquel Configuration

The classical projective planes are characterized by the fact that all possible Desargues configurations contained in them close. Similarly, the classical Möbius planes, of which the geometry of circles on the sphere is a prime example, are characterized by the fact that the so-called *Miquel configurations* close. For a Miquel configuration to close means that if you draw a cube on the sphere and the 5 sets of vertices of 5 of its faces are contained in one circle each, then the vertices of the remaining face are also all contained in one circle.

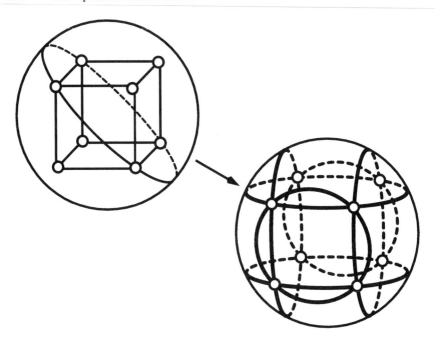

Note that the Miquel configuration is not an abstract plane configuration (see Chapter 3) as is the Desargues configuration, since some of the circles in the configuration intersect in two points.

Here is a flat picture of the Miquel configuration.

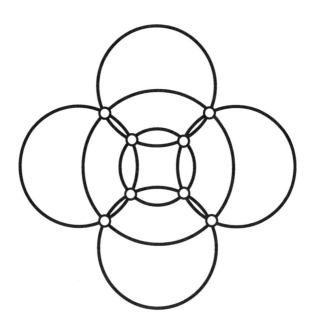

The Bundle Configuration

The ovoidal flat Möbius planes are characterized by the fact that the so-called *bundle configurations* close. For a bundle configuration to close means that if you draw a cube on the sphere and 5 of the following circles are present, then the sixth is also present. Note that four of the circles correspond to faces of the cube and two of them to diagonal planes. Below is a flat picture of the bundle configuration.

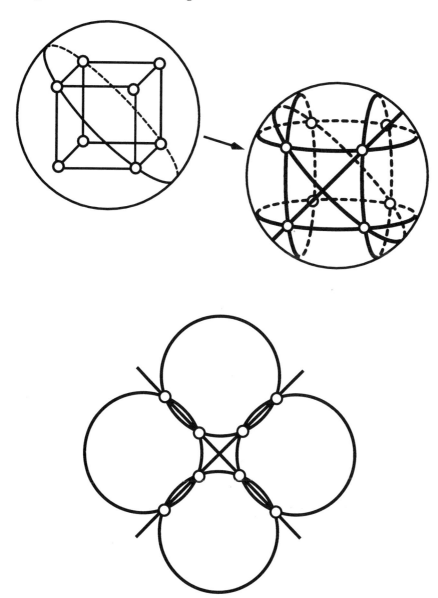

19.3   Nonclassical Flat Spherical Circle Planes

19.3.1   The Affine Part

Most constructions of nonclassical Möbius planes are variations of the geometry of Euclidean lines and circles in the xy-plane. Consider a geometry of topological circles and topological lines in the xy-plane. Via the inverse of the stereographic projection, this geometry turns into a geometry of topological lines and circles on the unit sphere. By extending all the topological lines by the north pole of the unit sphere, we arrive at a geometry of topological circles on the sphere. In order for this geometry to be a spherical circle plane, the original geometry on the plane has to satisfy the axiom of joining for spherical circle planes, plus the geometry of all topological lines in this plane has to be an $\mathbf{R}^2$-plane. If this is the case, we call the original geometry an affine part of the spherical circle plane, and we call the north pole of the sphere its *point at infinity*. In order for the second plane to be a flat Möbius plane, the first geometry also has to satisfy the axiom of touching for flat Möbius planes, plus the geometry of all topological lines in this plane has to be a flat affine plane.

Here are some constructions of affine parts of nonclassical flat Möbius planes.

19.3.2   Ewald's Affine Parts of Flat Möbius Planes

The affine part of the classical flat Möbius plane is the geometry of Euclidean lines and circles in the xy-plane. All of Ewald's affine parts also contain all the Euclidean lines, and it is only the set of topological circles that gets modified (see [33]).

Type 1: Let O be an arbitrary differentiable strictly convex curve in the plane, that is, an arbitrary topological oval in the Euclidean plane. Then the set of topological circles in this first affine part is generated by this curve alone. The generation rule is, "Shrink and inflate the curve O in all possible ways and translate the resulting curves in all possible ways." Note that if we start with the unit circle, we generate all Euclidean circles in this manner.

The following diagram shows an example of such a topological oval together with a selection of its images under inflation and shrinking.

The following diagram shows two typical pencils of circles in this affine part.

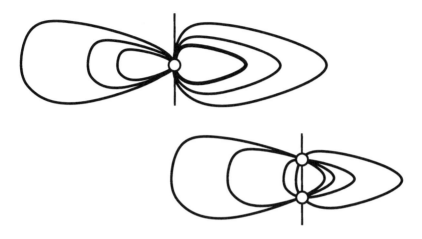

Type 2: Choose $t \geq 0$. Given a topological oval, we construct its *outer parallel curve at distance t*. It is the outer envelope of all Euclidean circles of radius t with centers on the topological oval.

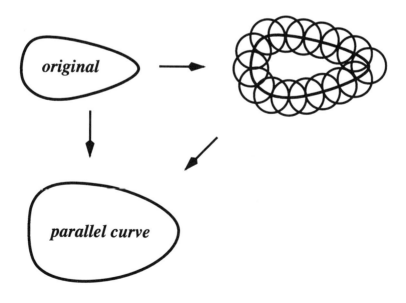

Note that the parallel curve of a shrunken version of the oval O will almost look like a circle if its diameter is "small" compared to t. On the other hand, the parallel curve of an expanded version of O will almost look like another expanded version of O if t is small compared with the diameter of the expanded version.

The set of topological circles in this second affine part consists of the Euclidean circles of radius less than or equal to t and the outer parallel curves at distance t of all topological ovals in the Type 1 construction. Clearly, all the translations of the xy-plane are automorphisms of the associated flat Möbius plane.

If we denote the affine part associated with the number t by A_t, then A_0 is just the affine part of Type 1. As we let t go from 0 to infinity, we see A being continuously deformed into the classical affine part. We call this deformation a *bubble isotopy*.

19.3.3   Steinke's Semiclassical Affine Parts

Steinke [98] constructs flat Möbius planes by cutting the classical flat Möbius plane in half along its equator and then fitting the two halves together again in a different way along the equator. This construction is very similar to the construction of the (flat affine) Moulton planes, and we will not go into any details here.

19.4   Classification

The flat Möbius planes with group dimension at least 3 have been classified by Strambach (see [107], [108], [109], and [110]). It turns out that the only flat Möbius plane with group dimension at least 4 is the classical plane; its group dimension is 6. The flat Möbius planes with group dimension 3 are Ewald's planes of Type 1 and a family that contains some semiclassical planes. A complete classification of all spherical circle planes of group dimension 2 should still be possible. As in the case of flat projective planes, it is hopeless to try to classify the flat Möbius planes with group dimension at least 1. Many of the ovoidal spherical circle planes are rigid; that is, they do not admit any automorphisms.

19.5   Subgeometries

The central part of the following map shows a spherical circle plane and its affine part. The icons surrounding the center stand for various geometries associated with it.

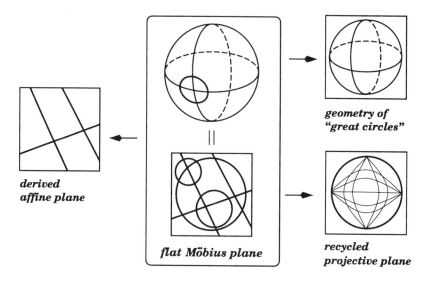

derived
affine plane

flat Möbius plane

geometry of
"great circles"

recycled
projective plane

The one on the left is just the derived geometry at a point, that is, an
$\mathbf{R}^2$-plane or a flat affine plane in the case of a flat Möbius plane.

19.5.1 Double Covers of Flat Projective Planes

The geometry of great circles, that is, the double cover of the classical flat
projective plane, is a subgeometry of the classical flat Möbius plane. This
geometry corresponds to the icon at the top right of the map. Subgeometries
like this also occur in nonclassical spherical circle planes. Note that the
geometry of great circles is the geometry of all those circles on the sphere
that are fixed by the antipodal map. Given a spherical circle plane, let i
be a fixed-point-free orientation-reversing homeomorphism of the sphere
to itself with the following property: If a point and its image under i are
both contained in a circle, then the circle is globally fixed by i. In this case
the geometry of all circles fixed by i is a double cover of a flat projective
plane. Here is an example. Start with an ovoidal spherical circle plane
represented, as usual, as the geometry of plane sections of some strictly
convex topological sphere. Then the set of all sections of the sphere with
planes containing a fixed inner point of the topological sphere is a double
cover of the classical projective plane. This double cover is the fix-geometry
of the bundle involution associated with the point.

19.5.2 Recycled Projective Planes

In Section 18.2 we gave a construction of disk models of flat projective
planes on the unit disk. Basically, this construction is a construction in
the classical flat Möbius plane, and it can be generalized to a construction

involving a general spherical circle plane. Let c be a circle in such a plane, let D be one of the two topological disks bounded by it, and let i be a fixed-point-free orientation-reversing involutory homeomorphism of c to itself. The points of a new geometry are the points in the interior of the disk plus all pairs of points that get exchanged by i. Its lines are c plus all intersections with the disk D of circles containing pairs of points that get exchanged by i.

See [77] for more information about this construction.

19.6   Lie Geometries Associated with Flat Möbius Planes

The Lie geometries associated with a flat Möbius plane are abstract geometries whose point and line sets are made up of points and circles of the Möbius plane and whose incidence relation is based on the touching of circles in the Möbius plane. We construct two examples of such geometries: Laguerre planes and divisible semibiplanes.

19.6.1   The Apollonius Problem

An important tool in verifying that the Lie geometries satisfy certain axioms is the solution of the so-called *Apollonius problem* for flat Möbius planes. Originally, the problem posed and solved by Apollonius of Perga reads as follows:

Given three things, each of which may be a point, a straight line, or a circle, to draw a circle that shall pass through the point or points (if such are given) and touch the straight lines or circles, as the case may be.

This, of course, is a problem in the classical flat Möbius plane, or equivalently, its affine part. In arbitrary flat Möbius planes it makes no sense to ask for the construction of common touching circles. It still makes sense to ask for the number of touching circles depending on the relative positions of the three objects we started with. Here is an example with three circles that have two common touching circles.

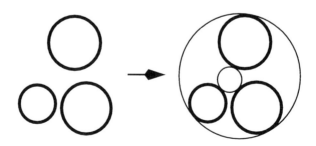

It turns out that if we draw circles in the respective relative position in a nonclassical flat Möbius plane, then the number and relative position of the touching circles will be the same as in the classical case. Here is a configuration of three circles in one of Ewald's planes that corresponds to the one above.

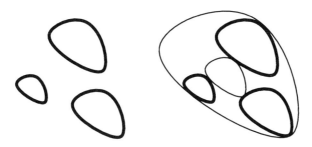

Here are two more examples of Apollonius configurations and common touching circles in the classical affine part.

See [93] for the complete solution of the Apollonius problem in topological Benz planes like the flat Möbius, Laguerre, and Minkowski planes. For the solution of the problem in the classical planes see [28] and [34].

19.6.2 Semibiplanes

Let c be a circle in a flat Möbius plane and let D be one of the two topological disks bounded by it. We define an abstract geometry as follows: Let both its Points and its Lines be the circles contained in D that touch c and let a Point be incident with a Line if and only if they touch in a point not contained in c. This geometry is a divisible semibiplane. The following diagram shows two Points being incident with exactly two Lines. This is followed by generator-only pictures of the Points and Lines of the semibiplane. These generator-only pictures also correspond to parallel classes of Points and Lines.

There are basically three different kinds of Apollonius configurations that we have to look at when checking the axioms for semibiplanes (see Chapter 11). Here they are.

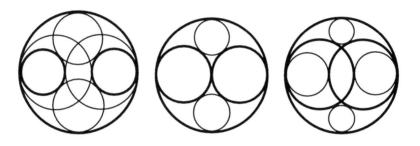

See [76], [82], and [83] for more information about this and related constructions of semibiplanes.

19.6.3   The Geometry of Oriented Lines and Circles

Construct a geometry from an affine part of a flat Möbius plane as follows: The Points are the oriented topological lines in the affine part. This means that there are two Points of the new geometry associated with any of the topological lines corresponding to the two possible orientations. Two Points are called parallel if they are parallel as topological lines and share the same orientation. The Circles of the new geometry are the oriented circles in the affine part of the Möbius plane plus the points of the xy-plane. A Point is incident with a Circle if the Circle is a point contained in the Point, or if the Point and the Circle touch and have the same orientation as depicted in the following diagram.

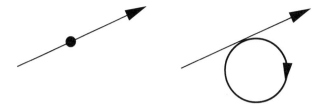

This incidence structure is an abstract Laguerre plane, that is, it satisfies axioms B1, B2, B3, and B5 (see Chapter 12), which can be checked using the solution of the Apollonius problem:

(B1) Any three pairwise nonparallel Points are contained in a unique Circle. (Here are the essentially different situations that occur.)

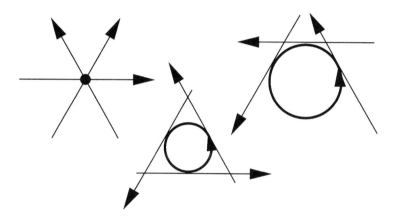

(B2) Given a Point p incident with a Circle c and a Point q not incident with the Circle, there is a unique Circle d that touches c in the geometrical sense and is incident with both p and q.

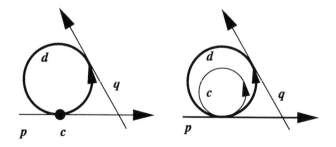

(B3) Given a Point p and a Circle c, there is exactly one Point q parallel to the Point incident with the Circle.

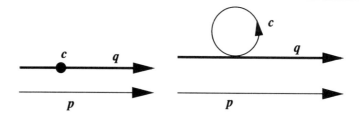

(B4) Every Circle contains at least 3 Points. Clearly, this is the case.

The abstract Laguerre planes arising in this manner can actually be shown to be isomorphic to flat Laguerre planes, which we are going to consider in the next chapter. The abstract Laguerre plane arising from the classical affine part is the classical Laguerre plane. For more information about this construction see [93], [36], and [37]. See also [4] for the classical case.

20
Cylindrical Circle Planes of Rank 3

20.1 Intro via Ovoidal Cylindrical Circle Planes

Cylindrical circle planes (of rank 3) behave very similarly to the spherical circle planes that we just considered and the toroidal circle planes that we are going to consider in the next chapter. Still, in some respects they are more complicated geometries, since we have to start worrying about different points being parallel. On the other hand, they have a much simpler structure, since the circle set of a cylindrical circle plane can be described in terms of continuous periodic functions. Also, the cylindrical circle planes are the only ones among the rank 3 circle planes that have natural higher-rank counterparts.

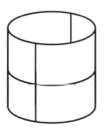

Just to reiterate, a *cylindrical circle plane* is a geometry whose point set is homeomorphic to the cylinder; two points are *parallel* if they are contained in the same vertical on the cylinder, and circles are topological circles that intersect every one of the verticals exactly once. The verticals are also called *parallel classes* (of points). The diagram above shows a sample parallel class and a typical circle.

The axiom of joining that every cylindrical circle plane satisfies guarantees a unique circle that contains three given *pairwise nonparallel* points. The derived geometry at a point of a cylindrical circle plane together with the verticals not containing the point is an $\mathbf{R}^2$-plane (check this!). This geometry is also called the *derived $\mathbf{R}^2$-plane*. A cylindrical circle plane is called a *flat Laguerre plane* if the derived $\mathbf{R}^2$-planes at all its points are flat affine planes. It is easy to show that a flat cylindrical circle plane is a flat Laguerre plane if and only if it satisfies the axiom of touching.

The classical example of a flat Laguerre plane is the geometry of nontrivial plane sections of the cylinder over the unit circle in the xy-plane. In general, given a strictly convex topological circle in the xy-plane, the geometry of plane sections of the cylinder over this curve is a cylindrical circle plane. These kinds of cylindrical circle planes are also called *ovoidal cylindrical circle planes*. An ovoidal cylindrical circle plane is a flat Laguerre plane only if the topological circle is differentiable, that is, if it is a topological oval in the Euclidean plane. If it has a "corner," the derived geometry at one of the points on the cylinder over this corner is no longer an affine plane, and the cylindrical circle plane is not a flat Laguerre plane.

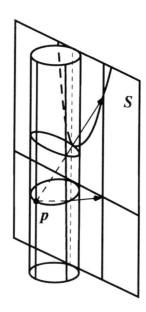

Consider a cylindrical circle plane on the cylinder as shown above. Let S be a tangent plane of the cylinder and p a point of the cylinder at maximal distance from this tangent plane. We stereographically project the cylinder from p onto this plane. This map is a homeomorphism of the cylinder minus the vertical v_p through p onto the plane. Under this stereographic projection, all circles of the circle plane are mapped onto topological lines of the tangent plane, and the verticals on the cylinder are mapped onto the verticals of the tangent plane. If the cylindrical circle plane under discussion is the classical example, then circles through the point p are mapped onto nonvertical Euclidean lines in the plane. For any other point on the vertical v_p, there is a constant $k \neq 0$ such that the images of circles through this point are parabolas $x \mapsto kx^2 + mx + n$, $m, n \in \mathbf{R}$. Of course, the set of all such parabolas plus the verticals in the tangent plane forms one of the parabola models of the Euclidean plane (see Section 16.3.2).

This shows that the derived $\mathbf{R}^2$-plane of this classical example at any of its points is the Euclidean plane. Convince yourself, using similar stereographic projections, that the derived geometries at all points of an ovoidal cylindrical circle plane are geometries that are restrictions of the Euclidean plane to an open vertical strip and that this subset is the full plane if the circle plane is a flat Laguerre plane.

20.2   The Miquel and Bundle Configurations

The Miquel Configuration

Like the classical Möbius plane, the classical Laguerre plane is characterized by the fact that certain *Miquel configurations* close. For a Miquel configuration in a cylindrical circle plane to close means that if you draw a cube on the cylinder such that no two vertices of a face are parallel and the 5 sets of vertices of 5 of the faces are contained in one circle each, then the vertices of the remaining face are also all contained in one circle.

Here is a picture of a Miquel configuration drawn in the classical affine part. The picture on the right shows the corresponding cube.

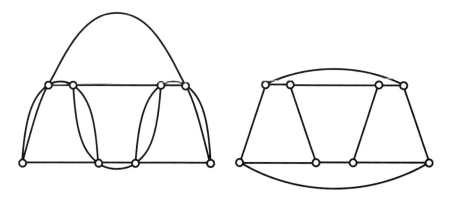

We already mentioned in the last chapter that the geometry of oriented Euclidean lines and circles is isomorphic to the classical flat Laguerre plane. Here is what a Miquel configuration looks like in this special kind of representation of this plane. Remember that the eight oriented Euclidean lines are points of the Laguerre plane, whereas the six oriented circles are circles of the Laguerre plane.

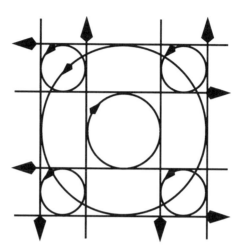

Note again that the Miquel configuration is not an abstract plane configuration, since some of the circles in the configuration intersect in two points.

The Bundle Configuration

The ovoidal flat Laguerre planes are characterized by the fact that the *bundle configurations* close. The bundle configurations are defined as in the case of spherical circle planes, with the only difference that only those configurations of four faces and two diagonal planes are of interest in which no two vertices in a face or a diagonal plane are parallel.

We modify the above Miquel configuration to arrive at a picture of a bundle configuration.

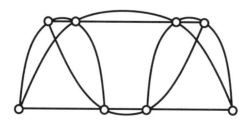

20.3   Nonclassical Flat Cylindrical Circle Planes

Unless the oval associated with an ovoidal cylindrical circle plane is a Euclidean ellipse, the circle plane will be nonclassical. In the following we give some more appealing constructions of nonclassical cylindrical circle planes.

20.3.1   Mäurer's Construction

In this section we want to recall some constructions of nonclassical flat Laguerre planes that are modifications of the construction of the classical Laguerre plane in Euclidean space.

We say it one more time: The classical flat Laguerre plane is obtained as the geometry of nontrivial plane sections of a cylinder in Euclidean space with a circle in the xy-plane as base.

We arrived at the ovoidal planes by replacing the circle in this construction by an arbitrary topological oval in the xy-plane.

Mäurer constructed nonclassical Laguerre planes by replacing the nonvertical planes that intersect the cylinder in the classical model by certain bent planes [69].

Let the circle in the classical construction be a circle in the xy-plane with center at the origin. Let A be an arbitrary vertical plane parallel to the xz-plane. Then every nonvertical plane E in Euclidean space intersects A and the yz-plane in one line each. Let l be the first line and let p be the infinite point of the second line. Then E can be considered as the set of all points in $\mathbf{R}^3$ that get projected on l through p (see the diagram on the left).

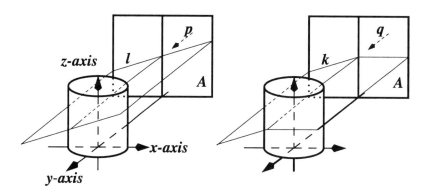

The plane A considered as a geometry is, of course, just a copy of the Euclidean plane. We now replace the nonvertical lines in A by lines that are bent at the yz-plane, such that the resulting incidence structure is a Moulton plane (see Section 16.3.1). Let k be a (bent) line in this plane and let q be an infinite point of one of the lines in the yz-plane. Then the set of all points in Euclidean space that get projected from q onto k is a "bent" plane (see the above diagram on the right). Let B be the set of all such bent planes. Now, the geometry of sections of these bent planes with the cylinder is (in general) a nonclassical Laguerre plane.

This construction has been generalized further by Hartmann [46], [47]; Kleinewillinghöfer [63]; and Steinke [103].

20.3.2 The Affine Part

Most constructions of nonclassical Laguerre planes are variations of the geometry of nonvertical Euclidean lines and parabolas with a vertical axis of symmetry. Consider a geometry of topological lines in the xy-plane that can be partitioned into flat affine planes and whose lines all intersect every single one of the verticals in the plane once. If this geometry, or one of its images under a homeomorphism of the plane that leaves the verticals invariant, arises from a cylindrical circle plane via the above stereographic projection, we call it an affine part. If it exists, this cylindrical circle plane is uniquely determined up to topological equivalence (this is a nontrivial result! See [85]). We call the vertical v_p on the cylinder the *vertical at infinity*.

Here are some constructions of affine parts of nonclassical flat Laguerre planes.

20.3.3 The Artzy–Groh Construction

Let P be the graph of a differentiable function $\mathbf{R} \to \mathbf{R}$ whose derivative is a homeomorphism of $\mathbf{R}$ to itself. We have already seen in the section on nonclassical flat affine planes that the translates of such a curve plus the verticals in the Euclidean plane form a flat affine (shift) plane. Shrinking and inflating P by a nonzero (possibly negative) factor gives another curve that is again a graph of a differentiable function whose derivative is a homeomorphism.

The following construction (see [1]) mimics the first of Ewald's constructions of flat Möbius planes that we described in the previous chapter. The set of topological circles in our affine part are the nonvertical Euclidean lines plus all the curves generated by the following rule: "Shrink and inflate the curve P in all possible ways and translate the resulting curves in all possible ways." Note that if we start with a Euclidean parabola having a vertical axis of symmetry, we generate all Euclidean parabolas having vertical axes of symmetry in this manner.

The following diagram shows an example of a suitable curve P together with a selection of its images under inflation and shrinking. This is followed by a diagram of two typical pencils of circles in this affine part.

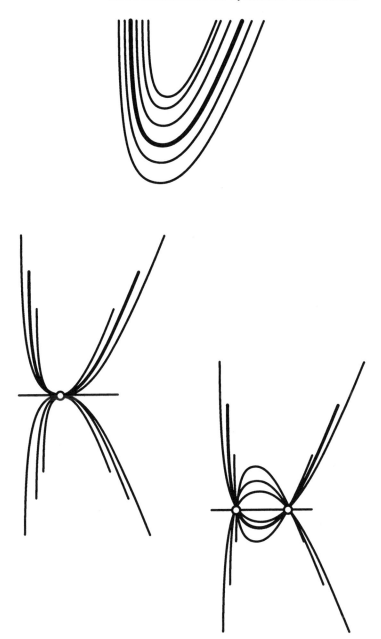

20.3.4   The Löwen–Pfüller Construction

Let P be the curve in the above construction. Then it is sometimes possible
to generate another affine part of a flat Laguerre plane from P with the

following generating rule: "Shrink and inflate P in the vertical direction only in all possible ways and translate all curves generated in this manner in all possible ways." See [67] and [68] for the precise conditions under which a set of curves generated like this, together with the nonvertical Euclidean lines, is an affine part of a flat Laguerre plane.

20.3.5   Steinke's Two Types of Semiclassical Planes

Steinke (see [99] and [100]) constructs flat Laguerre planes by cutting the classical flat Laguerre plane in half and then fitting the two halves together in a different way. He considers two different ways of cutting. First along the unit circle and second along two parallel classes. These constructions are very similar to the construction of the (flat affine) Moulton planes and Steinke's construction of semiclassical Möbius planes.

20.3.6   Integrated Flat Affine Planes

In order to understand this section you have to review Section 16.3.4. Consider a flat affine plane that contains all verticals in the xy-plane as lines. Then the verticals form one parallel class. Any other parallel class can be integrated to give an $\mathbf{R}^2$-plane, or if we are lucky, even a flat affine plane. It turns out that if the integral of every (nonvertical) parallel class is a flat affine plane, then the integral of the whole plane is the affine part of a flat Laguerre plane. See [75] for details about this construction.

20.4   Classification

The flat Laguerre planes with group dimension at least 5 have been classified by Löwen and Pfüller (see [67] and [68]). It turns out that the only flat Laguerre plane with group dimension at least 6 is the classical plane; its group dimension is 7. The flat Laguerre planes with group dimension 5 are some of the planes constructed by Löwen and Pfüller. A complete classification of all cylindrical circle planes of group dimensions 3 and 4 should still be possible. Many of the semiclassical planes are rigid; that is, they do not admit any automorphisms.

20.5   Subgeometries

The central part of the following map shows a cylindrical circle plane and its affine part. The icons surrounding the center stand for various geometries associated with it.

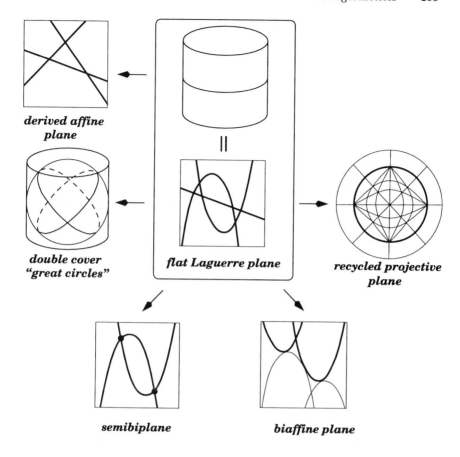

derived affine plane

double cover "great circles"

flat Laguerre plane

recycled projective plane

semibiplane

biaffine plane

The first one on the left is just the derived geometry at a point, that is, an $\mathbf{R}^2$-plane.

20.5.1 Recycled Projective Planes

Just as with gluing, there are two different ways of "recycling" cylindrical circle planes into flat projective planes, depending on whether we "recycle along" two parallel classes or along a circle.

Recycling Along Two Parallel Classes

Recycling along two parallel classes gives a representation of a flat projective plane on the Möbius strip. Consider the strip $[-1, 1] \times \mathbf{R}$ in an affine part of the cylindrical circle plane and identify the two verticals that bound the strip by an orientation-reversing homeomorphism like the one indicated in the following diagram. The lines of our geometry are verticals in the strip plus the intersections with the strip of all those circles whose points of intersection with the two bounding verticals get identified by the homeomorphism. The diagram on the right shows the pencil of lines through a pair of points that get identified.

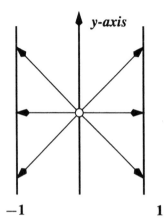
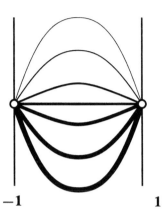

If we physically glue together the bounding verticals, as described by the homeomorphism, we get a flat projective plane minus a point represented on the Möbius strip.

Recycling Along a Circle

Recycling along a circle gives a representation of a flat projective plane on a disk. We concentrate on the restriction of the cylindrical circle plane to the cylinder that lies above a circle c in the plane. In real three-dimensional projective space the cylinder is really a cone over the unit circle with the point at infinity of the verticals in the space as vertex. Let us add this point to the upper cylinder. The resulting space is a topological disk.

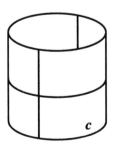

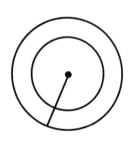

cylinder viewed as a cone

cone viewed from from above

Now choose a fixed-point-free orientation-reversing involutory homeo-morphism i of the circle. In the example depicted above we choose this homeomorphism to be the antipodal map of the circle. The points of the flat projective plane are the interior points of the topological disk (this includes the point at infinity) plus all pairs of points of the circle c that get exchanged by the homeomorphism i. We first construct the lines through the point at infinity. The homeomorphism i induces, in a natural way, an involution i' of the set of verticals on the cylinder. A line through the point at infinity is made up of the point itself and two verticals that get ex-changed by the involution i'. In our example these lines are simply the line segments connecting antipodal points on the circle. The remaining lines of the projective plane are c plus the intersections of the circles with the disk whose two points of intersection with the circle c get exchanged by the involution i. See [77] for more information about these constructions.

20.5.2   Semibiplanes

Let c be a circle in a cylindrical circle plane and let D be one of the two open cylinders bounded by it. We define a geometry as follows: Let its points be the points of D and its lines be the circles contained in $D \cup c$ that touch c. In the case of the classical Laguerre plane, it is very easy to see that this geometry is a divisible semibiplane, with points being parallel in the semibiplane if and only if they are parallel in the Laguerre plane, and lines being parallel if and only if they touch as circles in a point of c.

See [76], [82], and [83] for more information about this and related con-structions of semibiplanes.

20.6   Lie Geometries Associated with Flat Laguerre Planes

Similar to the Lie geometries associated with a flat Möbius plane, the Lie geometries associated with a flat Laguerre plane are abstract geometries whose point sets are made up of points and circles of the Laguerre plane and whose incidence relation is based on the touching of circles in the Laguerre plane. We construct examples for such geometries. Again the most important tool in verifying that the Lie geometries satisfy certain axioms is the solution of the *Apollonius problem* for flat Laguerre planes.

20.6.1   Biaffine Planes

Consider an affine part of a Laguerre plane and fix two points on the ver-tical at infinity. Construct an abstract geometry as follows: Its Points are the topological lines of the affine part associated with the first fixed point, and its Lines are the topological lines of the affine part associated with the second fixed point. For example, in the classical affine part the Points might correspond to all parabolas with leading coefficient 1, and the Lines might correspond to all the parabolas with leading coefficient -1. A Point is incident with a Line if the circles that correspond to them touch. Two Points in this geometry are called parallel if they are not both incident with a Line. Similarly, two Lines in this geometry are called parallel if they are not both incident with a Point. The parallel relations are equivalence relations, and the parallel classes of Points and Lines correspond to parallel classes of lines in the two derived affine planes at the fixed points, respectively.

This geometry is called a *biaffine plane*, and it satisfies the following two axioms:

Axioms for biaffine planes

(**Bi1**) Two nonparallel Points are incident with exactly 1 Line, and two nonparallel Lines are incident with exactly 1 Point.

(**Bi2**) Given a Line l and a Point p not incident with the Line, there is exactly one Line parallel to l that is incident with the Point. Furthermore, there is exactly one Point parallel to p incident with l.

The following diagram shows a parallel class of Points, a parallel class of Lines, and a Line pencil.

Here are two nonparallel Lines and the unique Point incident with both of them.

We remark that in the classical case depicted in the diagrams, the resulting geometry is isomorphic to the Euclidean plane with the verticals removed.

20.6.2 More Semibiplanes

The following construction of semibiplanes (see [84]) is closely related to the construction of biaffine planes in the previous section. Consider an affine part of a flat Laguerre plane and fix three points p, q, and r on the vertical at infinity such that q separates p and r. We construct a semibiplane as follows: Its Points are the points of the affine part plus the topological lines of the affine part associated with the point q. Its Lines are the topological lines associated with p and r. A Point of the geometry is incident with a Line if the Point is contained in the Line or if the Point and the Line touch.

For example, in the classical affine part the points of the xy-plane plus the nonvertical Euclidean lines may be taken as the Points and the Euclidean parabolas with leading coefficients -1 and 1 as the Lines. The following diagram shows the various ways in which two nonparallel Points in this semibiplane are incident with two nonparallel Lines.

20.6.3   Generalized Quadrangles

Generalized quadrangles can be constructed from flat Laguerre planes as follows: Let the Points of the geometry be the points of the cylinder plus all circles of the Laguerre plane plus one additional point O. An extended tangent pencil is a tangent pencil of the Laguerre plane plus the carrier of the pencil. An extended parallel class is a vertical on the cylinder to which the additional point O has been added. The Lines of the geometry are the extended parallel classes plus the extended tangent pencils.

For details about this construction see [93] and [34].

21
Toroidal Circle Planes

21.1 Intro via the Classical Minkowski Plane

In this chapter we concentrate on toroidal circle planes, the third kind of flat circle planes of rank 3.

A *toroidal circle plane* is a geometry whose point set is homeomorphic to the torus; two points are *parallel* if they are contained in the same vertical or horizontal circle on the torus, and circles are topological circles that intersect every one of the verticals and horizontals exactly once. The horizontal and vertical circles are also called the *parallel classes* of the plane. The following diagram shows a vertical, a horizontal, and a sample circle.

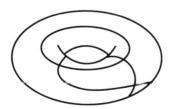

If we consider the torus to be the topological product of two unit circles $\mathbf{S}^1 \times \mathbf{S}^1$, then circles are just graphs of homeomorphisms from $\mathbf{S}^1$ to itself. Also, every toroidal circle plane satisfies an axiom of joining that is worded in exactly the same manner as that for cylindrical circle planes. There is a difference though. Points can be parallel in two different ways depending on whether they are both contained in a vertical or horizontal on the torus. If a toroidal circle plane satisfies the same axiom of touching as flat Laguerre planes, then we call it a *flat Minkowski plane*.

The classical example of such a circle plane is the geometry of plane sections of a hyperbolic quadric in real three-dimensional projective space. We remove from this three-dimensional projective space one of the planes that intersects the quadric in a circle. What we are left with is the Euclidean space, and the part of the quadric contained in this space is a hyperboloid. The verticals and horizontals on the torus correspond to Euclidean lines embedded in the hyperboloid.

There are no nonclassical ovoidal toroidal circle planes; that is, unlike the sphere and cylinder, the hyperbolic quadric cannot be deformed in a minor way to give toroidal circle planes.

In this chapter the *diagonals* of the Euclidean plane are the Euclidean lines with slope 1 or −1. A *topological hyperbola* in the Euclidean plane is the restriction of a topological oval in the real projective plane that intersects the line at infinity in the two infinite points of the diagonals. If the topological oval is a nondegenerate conic, then the topological hyperbola is a *Euclidean hyperbola*. All these hyperbolas have two perpendicular diagonals as asymptotes and intersect all other diagonals in exactly one point each. The following diagram shows some diagonals and two topological hyperbolas.

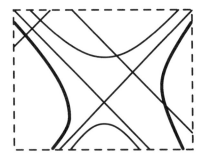

Consider a toroidal circle plane whose point set has been identified via a homeomorphism with a hyperbolic quadric such that parallel classes are identified with parallel classes. Let S be a tangent plane of the quadric at the point p. Then this plane contains two of the parallel classes that intersect in the point.

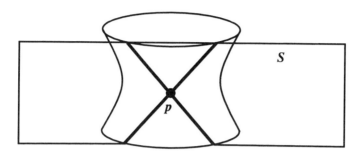

Choose a plane parallel to this tangent plane and stereographically project the circle plane minus the two parallel classes onto the plane. The parallel classes are mapped onto the diagonals of the plane, all circles through p are mapped onto Euclidean lines different from the diagonals, and all circles not through p are mapped onto topological hyperbolas.

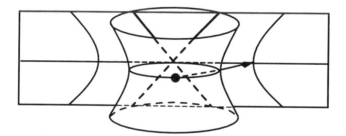

If we start with the classical plane, this projected geometry is the geometry on the xy-plane whose circles are the nondiagonal Euclidean lines and all Euclidean hyperbolas.

21.2 The Miquel, Bundle, and Rectangle Configurations

The Miquel and Bundle Configurations

Like the classical Möbius plane and the classical Laguerre plane, the classical Minkowski plane is also characterized among the flat Minkowski planes by the fact that all *Miquel configurations* close. Since the classical Minkow-ski plane is the only ovoidal flat Minkowski plane, it is also characterized by the fact that all *bundle configurations* close.

The Rectangle Configuration

The classical Minkowski plane is also characterized among the toroidal circle planes as the only such plane in which every *rectangle configuration* closes. For a rectangle configuration to close means that if we draw four rectangles as follows using only segments of parallel classes for sides of the rectangles and if three of the four sets of corresponding vertices of the rectangles are contained in circles of the circle plane, then the fourth set is also contained in a circle.

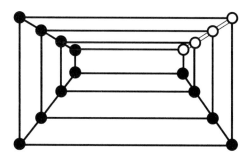

21.3   Nonclassical Toroidal Circle Planes

21.3.1   The Affine Part

Many constructions of nonclassical Minkowski planes are variations of the geometry of nondiagonal Euclidean lines and Euclidean hyperbolas in the xy-plane. Consider a geometry of topological lines and hyperbolas on the xy-plane such that every single one of the topological lines intersects every single one of the diagonals in exactly one point. Via the inverse of the above stereographic projection, this geometry turns into a geometry on the hyperbolic quadric. We turn this geometry into a geometry of topological circles on the quadric by (1) extending the image of every topological line by the point p; and (2) extending the image of every topological hyperbola by the two points on the two parallel classes through p that complete the images of the two asymptotes of the hyperbola. If the resulting geometry is a toroidal circle plane, we call the original geometry an affine part of the toroidal circle plane.

Here are some constructions of affine parts of nonclassical flat Minkowski planes.

21.3.2   The Two Parts of a Toroidal Circle Plane

We have already mentioned that circles in a toroidal circle plane T can be considered as the graphs of homeomorphisms of $\mathbf{S}^1$ to itself. Call the set of all those circles that correspond to orientation-preserving and orientation-reversing homeomorphisms the positive part and the negative part of T, respectively. These two sets are always nonempty and are really completely independent geometries in their own right. In fact, given two toroidal circle planes or flat Minkowski planes, combining the positive part of one with the negative part of the other gives a new toroidal circle plane or flat Minkowski plane, respectively. So far we know only one flat Minkowski plane. Still, we can use this construction to generate nonclassical Minkowski planes from this classical plane as follows: Just use a sufficiently "bad" homeomorphism of the torus to itself that preserves the parallel classes and combine the positive part of the original with the negative part of the same plane whose point set has been deformed by the homeomorphism.

Here is an example. Note that in an affine part of a toroidal circle plane one part corresponds to the set of all topological hyperbolas whose branches are open above and below and all Euclidean lines with slope greater than -1 and less than 1. The other part corresponds to the set of all other topological hyperbolas and lines in the affine part.

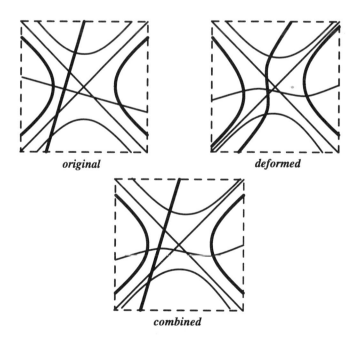

original   *deformed*

combined

See [79], [86], and [102] for more information about this construction.

21.3.3   The Artzy–Groh Construction

The construction of nonclassical flat Laguerre planes from a single topological parabola described in Section 20 has a counterpart for flat Minkowski planes. Let H be a topological hyperbola. Shrinking and inflating H by a positive factor gives another topological hyperbola. If we shrink and inflate with the point of intersection of the two asymptotes of the hyperbola as center, the images of the hyperbola partition the plane minus the asymptotes and the quarter planes the hyperbola is not contained in.

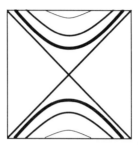

The images of all the hyperbolas in this partition under all possible translations together with all Euclidean lines with slope greater than -1 and less than 1 forms one-half of an affine part of a flat Minkowski plane. Of course, if we start with a Euclidean hyperbola, this will be one of the two parts of the classical plane.

21.3.4   Semiclassical Planes

Consider again the above partition of the Euclidean plane into topological hyperbolas. Take a branch of such a hyperbola that is open above together with an arbitrary branch of a hyperbola that is open below. This gives a new topological hyperbola. We modify the above partition by recombining the two different kinds of branches. This recombination can be done in an arbitrary way as long as the overall nested structure of the partition is preserved. The end result might look as follows.

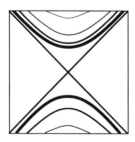

Now we continue to generate one-half of a flat Minkowski plane as in the Artzy–Groh construction; that is, the images of all the hyperbolas in this partition under all possible translations together with all Euclidean lines with slope greater than −1 and less than 1 form one-half of an affine part of a flat Minkowski plane.

See [48], [79], and [101] for more information about this construction.

21.3.5 Proper Toroidal Circle Planes

So far we have not seen any toroidal circle planes that are not flat Minkowski planes. Halves of such planes can be constructed by again modifying the Artzy–Groh construction. Start with the classical partition and collapse the shaded cross as indicated by the arrows in the following sequence of diagrams.

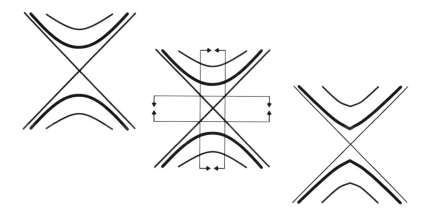

Now we continue to generate one-half of a proper toroidal circle plane as in the Artzy–Groh construction.

See [79] for more information about this construction.

21.4 Classification

The flat Minkowski planes with group dimension at least 4 have been classified by Schenkel (see [92]). It turns out that the only flat Minkowski plane with group dimension at least 5 is the classical plane; its group dimension is 6. The flat Minkowski planes with group dimension 4 are one particular class of semiclassical planes. A complete classification of all toroidal circle planes of group dimension 3 should still be possible. Many of the semiclassical planes are rigid; that is, they do not admit any automorphisms.

21.5   Subgeometries

The central part of the following map shows a toroidal circle plane and its affine part. The icons surrounding the center stand for various geometries associated with it.

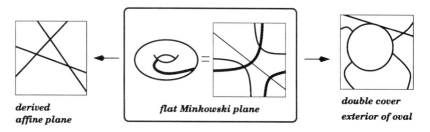

derived
affine plane

flat Minkowski plane

double cover
exterior of oval

21.5.1   Flat Projective Planes Minus Convex Disks

In the section on subgeometries of spherical circle planes we identified double covers of flat projective planes as fix-geometries of certain involutory homeomorphisms of the sphere. A similar construction is possible for flat toroidal planes.

Given a toroidal circle plane, let i be an involutory homeomorphism of the torus to itself that exchanges the horizontals with the verticals on the torus and that fixes exactly one circle pointwise. Furthermore, if a point and its image under i are both contained in a circle, then the circle is fixed by i. The geometry whose lines consist of all circles fixed by i plus all unions of parallel classes that get exchanged by i is a double cover of the exterior of a topological oval in a flat projective plane. We can always represent the toroidal circle plane such that in an affine part of it the circle fixed by i corresponds to the y-axis and the involution, when restricted to the xy-plane, is just the reflection of the xy-plane through the y-axis. Of course, the classical affine part admits such an involution. The following diagram is a generator-only picture of this geometry. The generating rule is, "Translate in the vertical direction." The diagram on the right shows what the different curves correspond to in the fix-geometry.

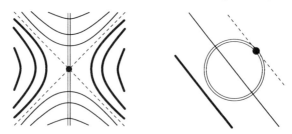

See [80] for more information about this kind of fix-geometry.

Appendix A
Models on Regular Solids

The following table is an index of the different models of geometries on the regular three- and four-dimensional solids that we consider in this book. The number on the left is the number of points of the respective geometry, and the number on the right is the number of the page on which we first introduce the model.

	Tetrahedron	
4	affine plane of order 2	9
	dual complete quadrangle	86
	biplane of order 1	136
6	complete quadrangle	35, 86
7	Fano plane	15
	biplane of order 2	139
10	inversive plane of order 3	163
	Desargues configuration	35, 124
	APC with parameters (10_3) (no PC)	35, 124
	Petersen graph	88
	Petersen system	89
11	biplane of order 3	142
15	generalized quadrangle of order $(2, 2)$	43, 87
	projective three-space of order 2	82
	an APC with parameters (15_3)	87
21	projective plane of order 4	113

		Cube/Octahedron	
	6	Laguerre plane of order 2	169
	8	one-point extension of the Fano plane	21
		generalized quadrangle of order $(1,4)$	41
		affine three-space of order 2	69
		semi-biplane of order $(8,4)$	157
	10	Petersen graph	52
	12	Schläfli's double six	62
	13	projective plane of order 3	103
	15	an APC with parameters (15_3)	47
		Icosahedron/Dodecahedron	
	6	the $2 - (6,3,2)$ design	20
	10	Petersen graph	51
	12	semibiplane of order $(12,5)$	158
	15	generalized quadrangle of order $(2,2)$	48
		projective three-space of order 2	90
	16	a biplane of order 4	148
	31	projective plane of order 5	120
		5-Cell	
	5	inversive plane of order 2	162
	10	Desargues configuration	122
		8-Cell = Tesseract	
	16	a biplane of order 4	149
		a semibiplane of order $(16,5)$	158
		16-Cell	
	8	extension of the Laguerre plane of order 2	170

Appendix B
Mirror Technique Stereograms

If you cannot see stereograms with either the parallel or cross-eyed technique, here is a complete list of the stereograms in this book drawn so that they can be viewed with the mirror technique.

B.1   The Generalized Quadrangle of Order $(4, 2)$

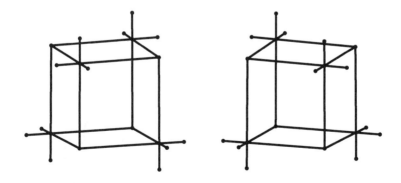

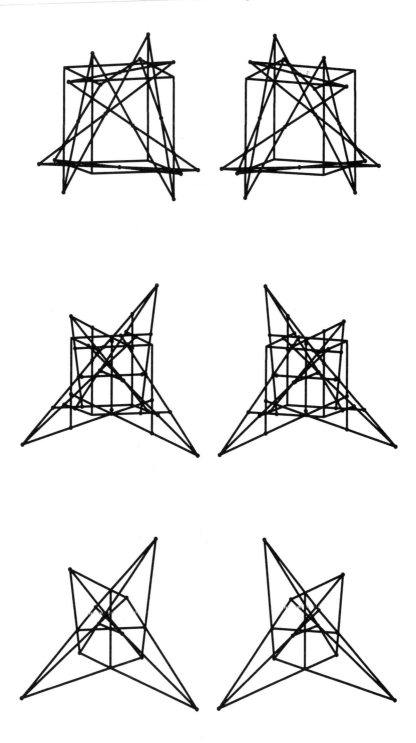

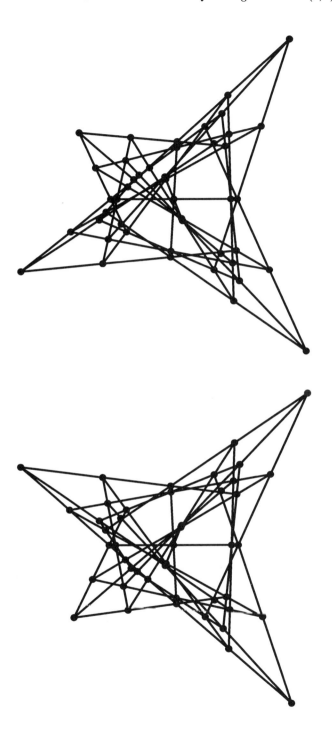

B.2   Two PCs with Parameters 10_3

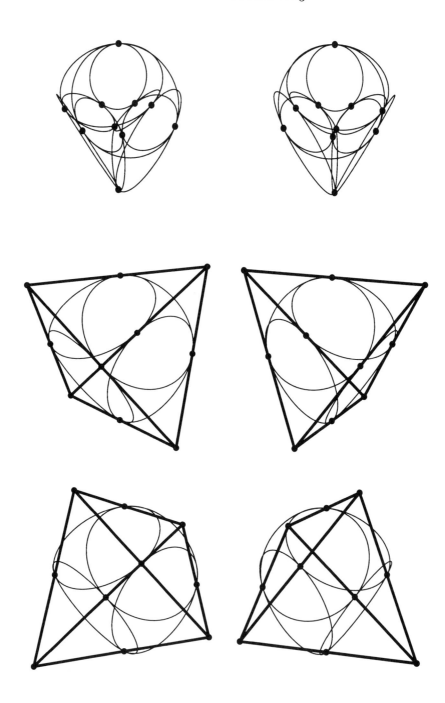

B.3   The Smallest Projective Space

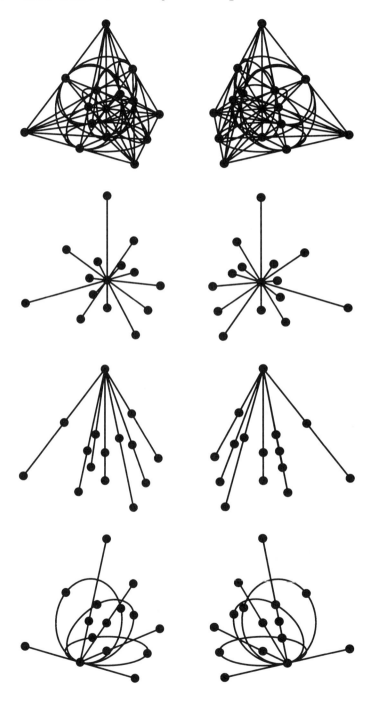

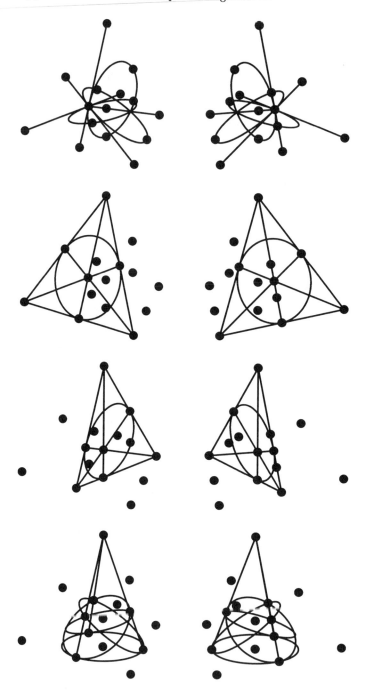

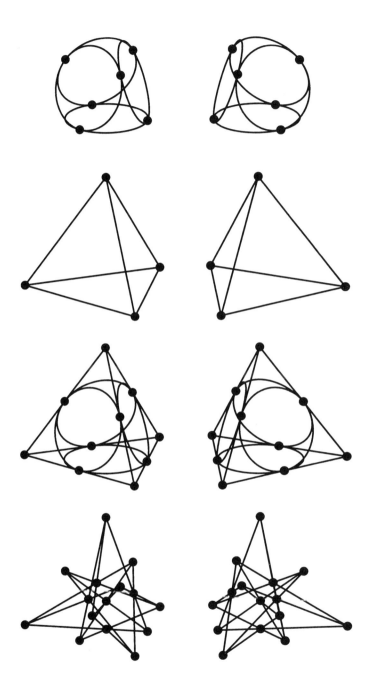

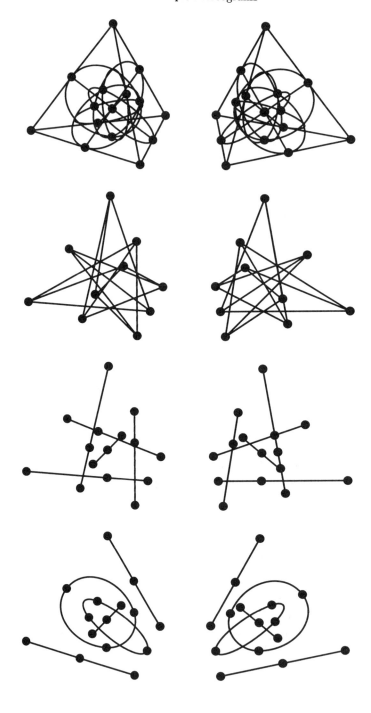

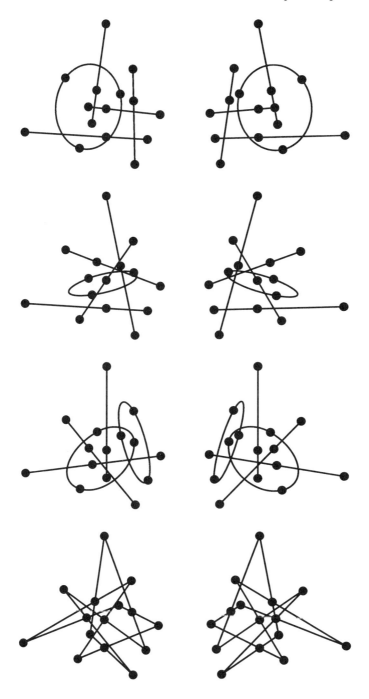

References

[1] Artzy, R. and Groh, H. Laguerre and Minkowski planes produced by dilatations. J. Geometry 26: 1–20, 1986.

[2] Batten, L. M. and Beutelspacher, A. The theory of finite linear spaces. Cambridge University Press, 1993.

[3] Baumert, L. D. Cyclic difference sets. Lecture Notes in Mathematics 182, Springer, 1971.

[4] Benz, W. Vorlesungen über Geometrie der Algebren. Springer, 1973.

[5] Bernardi, L. and Beutelspacher, A. Blocking sets of the known biplanes. Rend. Mat. Appl. (7) 7: 63–76, 1987.

[6] Bernardi, L. and Beutelspacher, A. Constructing the biplanes of order four from six points. Boll. Un. Mat. Ital. A (7) 2: 285–289, 1988.

[7] Beth, T., Jungnickel, D., and Lenz, H. Design theory. Cambridge University Press, 1986.

[8] Betten, A. and Betten, D. Tactical decompositions and some configurations n_4. Preprint.

[9] Betten, D. and Schumacher, U. The ten configurations 10_3. Rostock. Math. Kolloq. 46: 3–10, 1993.

[10] Beutelspacher, A. Einführung in die endliche Geometrie II. Bibliographisches Institut Mannheim/Wien/Zürich, 1983.

[11] Beutelspacher, A. $21 - 6 = 15$: a connection between two distinguished geometries. Amer. Math. Monthly 93: 29–41, 1986.

[12] Beutelspacher, A. A defense of the honour of an unjustly neglected little geometry or a combinatorial approach to the projective plane of order five. J. Geometry 30: 182–195, 1987.

[13] Bokowski, J. and Sturmfels, B. Computational synthetic geometry. Lecture Notes in Mathematics 1355, Springer, 1989.

[14] Bondy, J. A. and Murty, U. S. R. Graph theory with applications. Macmillan Press, 1976.

[15] Breach, D. R. Star gazing in affine planes, in: Combinatorial Mathematics IX, Lecture Notes in Mathematics 952, pp. 1–33, Springer, 1982.

[16] Brooke, M. Coin games and puzzles. Dover, 1973.

[17] Brouwer, A. E. Block designs, in: Handbook of combinatorics, vol. 1, P.L. Graham et al. eds., pp. 693–745, Elsevier, 1995.

[18] Brouwer, A. E., Cohen, A. M., and Neumaier, A. Distance-regular graphs. Ergebnisse der Mathematik und ihrer Grenzgebiete, 3. Folge, Band 18, Springer, 1989.

[19] Bruen, A. A. Kummer configurations and designs embedded in planes. J. Combin. Theory Ser. A 52: 154–157, 1989.

[20] Buchanan, T., Hähl, H., and Löwen, R. Topologische Ovale. Geom. Dedicata 9: 401–424, 1980.

[21] Buekenhout, F. ed. Handbook of incidence geometry. Elsevier, 1995.

[22] Cameron, P. J. Parallelisms of complete designs. London Mathematical Society Lecture Note Series 23, Cambridge University Press, 1976.

[23] Cameron, P. J. Combinatorics—topics, techniques, algorithms. Cambrige University Press, 1994.

[24] Cameron, P. J. and van Lint, J. H. Designs, graphs, codes and their links. London Mathematical Society Student Texts 22, Cambridge University Press, 1991.

[25] Colbourn, J. and Dinitz, J. H. eds. The CRC handbook of combinatorial designs. CRC Press Series on Discrete Mathematics and its Applications. CRC Press, Boca Raton, FL, 1996.

[26] Coxeter, H. S. M. Self-dual configurations and regular graphs. Bull. Amer. Math. Soc. 56: 413–455, 1950.

[27] Coxeter, H. S. M. Introduction to geometry. John Wiley and Sons, 1962.

[28] Coxeter, H. S. M. The problem of Apollonius. Amer. Math. Monthly 75: 5–15. 1968.

[29] Coxeter, H. S. M. Desargues configurations and their collineation groups. Math. Proc. Camb. Phil. Soc. 78: 227–246, 1975.

[30] Dembowski, P. Finite geometries. Ergebnisse der Mathematik und ihrer Grenzgebiete 44, Springer, 1968.

[31] Dorwart, H. L. The geometry of incidence. Prentice-Hall, 1966.

[32] Dudney, H. E. Amusements in mathematics. Dover, 1958.

[33] Ewald, G. Aus konvexen Kurven bestehende Möbiusebenen. Abh. Math. Sem. Univ. Hamburg 30: 179–187, 1967.

[34] Fisher, Ch. Models and theorems of the classical circle planes. Abh. Math. Sem. Univ. Hamburg 63: 245–264, 1993.

[35] Gardner, M. Mathematical carnival. Penguin, 1975.

[36] Groh, H. Laguerre planes generated by Moebius planes. Abh. Math. Sem. Univ. Hamburg 40: 43–63, 1974.

[37] Groh, H. Flat Moebius and Laguerre planes. Abh. Math. Sem. Univ. Hamburg 40: 65–76, 1974.

[38] Groh, H. $\mathbf{R}^2$-planes with 2-dimensional point transitive automorphism group, in: Proc. of the Lattice Theory Conference (Ulm, 1975), pp. 80–91, Univ. Ulm, 1975.

[39] Groh, H. Point homogeneous flat affine planes. J. Geometry 8: 145–162, 1976.

[40] Groh, H. Pasting of $\mathbf{R}^2$-planes. Geom. Dedicata 11: 69–98, 1981.

[41] Groh, H. Isomorphism types of arc planes. Abh. Math. Scm. Univ. Hamburg 52: 133–149, 1982.

[42] Grundhöfer, T. and Löwen, R. Linear topological geometries, in: Handbook of incidence geometry, F. Buekenhout, ed., pp. 1255–1324, Elsevier, 1995.

[43] Guy, R. K. The unity of combinatorics. Combinatorics advances (Tehran, 1994), Math. Appl. 329, pp. 129–159, Kluwer, 1995.

[44] Hall, J. I. On identifying PG(3, 2) and the complete 3-design on seven points. Ann. Discrete Math. 7: 131–141, 1980.

[45] Hall, M., Jr. Combinatorial theory, 2nd edition. Wiley-Interscience, 1986.

[46] Hartmann, E. Moulton-Laguerre-Ebenen. Arch. Math. 27: 424–435, 1976.

[47] Hartmann, E. Eine Klasse nicht einbettbarer Laguerre-Ebenen. J. Geometry 13: 49–67, 1979.

[48] Hartmann, E. Beispiele nicht einbettbarer reeller Minkowski-Ebenen. Geom. Dedicata 10: 155–159, 1981.

[49] Hilbert, D. and Cohn-Vossen, S. Anschauliche Geometrie, 2nd edition. Springer, 1996. English translation: Geometry and the imagination. Chelsea Publishing Company, 1952.

[50] Hirschfeld, J. W. P. Projective geometries over finite fields. Oxford University Press, 1979.

[51] Hirschfeld, J. W. P. Finite projective spaces of three dimensions. Oxford University Press, 1985.

[52] Hirschfeld, J. W. P. and Thas, J. A. General Galois geometries. Oxford University Press, 1991.

[53] Holton, D. A. and Sheehan, J. The Petersen graph. Australian Mathematical Society Lecture Series 7, Cambridge University Press, 1993.

[54] Horibuchi, S. ed. Stereogram. Cadence Books, San Francisco, 1994.

[55] Horibuchi, S. ed. Super stereogram. Cadence Books, San Francisco, 1994.

[56] Hughes, D. R. and Piper, F. C. Projective planes. Springer, 1973.

[57] Hughes, D. R. and Piper, F. C. Design theory. Cambridge University Press, 1985.

[58] Ito, N. and Leon, J. S. A Hadamard matrix of order 36. J. Combin. Theory Ser. A 34: 244–247, 1983.

[59] Jeurissen, R. H. The Petersen graph as a box of pandora. Nieuw Arch. Wisk. 3: 219–233, 1985.

[60] Jeurissen, R. H. Special sets of lines in PG(3, 2). Linear Algebra Appl. 226/228: 617–638, 1995.

[61] Kárteszi, F. Introduction to finite geometries. North Holland, 1976.

[62] Kempe, A. B. A memoir on the theory of mathematical form. Phil. Trans. Roy. Soc. London 177: 1–70, 1886.

[63] Kleinewillinghöfer, R. Eine Klassifikation der Laguerre-Ebenen. Dissertation, Darmstadt, 1979.

[64] Lehti, R. On the construction of elementary models for finite plane geometries. Treizième congrès des mathématiciens scandinaves, tenu à Helsinki, 18-23 août 1958, pp. 141-170, Mercators Tryckeri, Helsinki, 1958.

[65] Lemay, F. Le dodécaèdre et la géométrie projective d'ordre 5, in: Finite Geometries, Lecture Notes in Pure and Applied Mathematics 82, Johnson, N. et al. eds., pp. 279–306, Marcel Dekker, 1983.

[66] Lloyd, E. K. The reaction graph of the Fano plane, in: Combinatorics and graph theory '95, vol. 1, Ku Tung-Hsin, ed., pp. 260–274, World Scientific, 1995.

[67] Löwen, R. and Pfüller, U. Two-dimensional Laguerre planes over convex functions, Geom. Dedicata 23: 73–85, 1987.

[68] Löwen, R. and Pfüller, U. Two-dimensional Laguerre planes with large automorphism groups, Geom. Dedicata 23: 87–96, 1987.

[69] Mäurer, H. Eine Kennzeichnung halbovoidaler Laguerre-Geometrien. J. Reine Angew. Math. 253: 203–213, 1972.

[70] Moorhouse, G. E. Reconstructing projective planes from semibiplanes, in: Coding theory and design theory. Part II, IMA Vol. Math. Appl. 21: 280–285, 1990.

[71] Moulton, F. R. A simple non-desarguesian plane geometry. Trans. Amer. Math. Soc. 3: 192–195, 1902.

[72] Oxley, J. G. Matroid theory. Oxford University Press, 1992.

[73] Payne, S. E. Finite generalized quadrangles: a survey. Proceedings of the International Conference on Projective Planes (Washington State Univ., Pullman, Wash., 1973), pp. 219–261. Washington State Univ. Press, Pullman, Wash., 1973.

[74] Payne, S. E. and Thas, J. A. Finite generalized quadrangles. Research Notes in Math. 110, Pitman, Boston, 1984.

[75] Polster, B. Integrating and differentiating two-dimensional incidence structures. Arch. Math. 64: 75–85, 1995.

[76] Polster, B. Semi-biplanes on the cylinder. Geom. Dedicata 58: 145–160, 1995.

[77] Polster, B. Recycling circle planes. Bull. Austral. Math. Soc. 53: 325–340, 1996.

[78] Polster, B. N-unisolvent sets and flat incidence structures. Trans. Amer. Math. Soc., to appear.

[79] Polster, B. Toroidal circle planes that are not Minkowski planes. J. Geometry, to appear.

[80] Polster, B. Invertible sharply n-transitive sets. J. Combin. Theory Ser. A, to appear.

[81] Polster, B. Isotopy of topological planes. preprint.

[82] Polster, B. and Schroth, A. E. Semi-biplanes and antiregular generalised quadrangles. Geom. Dedicata, to appear.

[83] Polster, B. and Schroth, A. E. Plane models for semi-biplanes. Beiträge Algebra Geom., to appear.

[84] Polster, B. and Schroth, A. E. Split semi-biplanes in antiregular generalized quadrangles. Bull. Belg. Math. Soc., to appear.

[85] Polster, B. and Steinke, G. F. Criteria for two-dimensional circle planes. Beiträge Algebra Geom. 35: 181–191, 1994.

[86] Polster, B. and Steinke, G. F. Cut and paste in 2-dimensional projective planes and circle planes. Canad. Math. Bull. 38: 469–480, 1995.

[87] Polster, B. and Steinke, G. F. The inner and outer space of 2-dimensional Laguerre planes. J. Austral. Math. Soc. Ser. A 62: 104–127, 1997.

[88] Polster, B., Steinke, G. F. and Rosehr, N. On the existence of topological ovals in flat projective planes. Arch. Math. 68: 418–429, 1997.

[89] Salzmann, H. Topological planes. Adv. Math. 2: 1–60, 1967.

[90] Salzmann, H., Betten, D., Grundhöfer, T., Hähl, H., Löwen, R., and Stroppel, M. Compact projective planes. de Gruyter, 1995.

[91] Salwach, C. J. Planes, biplanes and their codes. Amer. Math. Monthly 88: 106–125, 1981.

[92] Schenkel, A. Topologische Minkowskiebenen. Dissertation, Universität Erlangen-Nürnberg, 1980.

[93] Schroth, A. E. Topological circle planes and topological quadrangles. Pitman Research Notes in Mathematics Series 337. Longman, Harlow, 1995.

[94] Schroth, A. E. How to draw a hexagon, preprint.

[95] Schroth, A. E. and Van Maldeghem, H. Half-regular and regular points in compact polygons. Geom. Dedicata 51: 215–233, 1994.

[96] Shaw, R. Icosahedral sets in PG(5, 2). Europ. J. Combinatorics 18: 315–339, 1997.

[97] Steinke, G. F. Topological affine planes composed of two Desarguesian halfplanes and projective planes with trivial collineation group. Arch. Math. 44: 472–480, 1985.

[98] Steinke, G. F. Semiclassical topological Möbius planes. Resultate Math. 9: 166–188, 1986.

[99] Steinke, G. F. Semiclassical topological flat Laguerre planes obtained by pasting along a circle. Resultate Math. 12: 207–221, 1987.

[100] Steinke, G. F. Semiclassical topological flat Laguerre planes obtained by pasting along two parallel classes. J. Geometry 32: 133–156, 1988.

[101] Steinke, G. F. 2-dimensional Minkowski planes and Desarguesian derived affine planes. Abh. Math. Sem. Univ. Hamburg 60: 61–69, 1990.

[102] Steinke, G. F. A family of 2-dimensional Minkowski planes with small automorphism groups. Resultate Math. 26: 131–142, 1994.

[103] Steinke, G. F. Topological circle geometries, in: Handbook of incidence geometry, F. Buekenhout, ed., pp. 1325–1354, Elsevier, 1995.

[104] Stevenson, F. W. Projective planes. W. H. Freeman and Co., San Francisco, 1972.

[105] Strambach, K. Über sphärische Möbiusebenen. Arch. Math. 18: 208–211, 1967.

[106] Strambach, K. Sphärische Kreisebenen. Math. Z. 113: 266–292, 1970.

[107] Strambach, K. Sphärische Kreisebenen mit dreidimensionaler nicht einfacher Automorphismengruppe. Math. Z. 124: 289–314, 1972.

[108] Strambach, K. Sphärische Kreisebenen mit einfacher Automorphismengruppe. Geom. Dedicata 1: 182–220, 1973.

[109] Strambach, K. Kollineationsgruppen sphärischer Kreisebenen. Abh. Math. Sem. Univ. Hamburg 41: 133–153, 1974.

[110] Strambach, K. Der Kreisraum einer sphärischen Möbiusebene. Monatsh. Math. 78: 156–163, 1974.

[111] Stroppel, M. A note on Hilbert and Beltrami systems. Resultate Math. 24: 342–347, 1993.

[112] Sturmfels, B. and White, N. All 11_3 and 12_3-configurations are rational. Aequationes Math. 39: 254–260, 1990.

[113] Van Dam, E. Classification of spreads of $PG(3,4)\backslash PG(3,2)$. Des. Codes Cryptogr. 3: 193–198, 1993.

[114] Van Maldeghem, H. Generalized polygons. Birkhäuser, to appear.

[115] Wild, P. On semibiplanes. Ph.D. thesis, Westfield College, University of London, 1980.

[116] Wild, P. Generalized Hussain graphs and semibiplane with $k \leq 6$. Ars Combin. 14: 147–167, 1982.

[117] Wild, P. Divisible semibiplanes and conics of Desarguesian biaffine planes. Simon Stevin 58: 153–166, 1984.

[118] Wild, P. Biaffine planes and divisible semiplanes. J. Geometry 25: 121–130, 1985.

[119] Wild, P. Some families of semibiplanes. Discrete Math. 138: 397–403, 1995.

[120] Wölk, R. D. Topologische Möbiusebenen. Math. Z. 93: 311–333, 1966.

[121] Wong, P. K. Cages—A survey. J. Graph Theory 6: 1–22, 1982.

[122] Yaglom, I. M. A simple non-Euclidean geometry and its physical basis. Springer, 1979.

Index

Universitext *(continued)*

Rotman: Galois Theory
Rubel/Colliander: Entire and Meromorphic Functions
Sagan: Space-Filling Curves
Samelson: Notes on Lie Algebras
Schiff: Normal Families
Shapiro: Composition Operators and Classical Function Theory
Simonnet: Measures and Probability
Smith: Power Series From a Computational Point of View
Smoryński: Self-Reference and Modal Logic
Stillwell: Geometry of Surfaces
Stroock: An Introduction to the Theory of Large Deviations
Sunder: An Invitation to von Neumann Algebras
Tondeur: Foliations on Riemannian Manifolds
Zong: Strange Phenomena in Convex and Discrete Geometry